Transistor Scaling—Methods, Materials and Modeling

MATERIALS RESEARCH SOCIETY
SYMPOSIUM PROCEEDINGS VOLUME 913

Transistor Scaling—Methods, Materials and Modeling

Symposium held April 18–19, 2006, San Francisco, California, U.S.A.

EDITORS:

Scott Thompson
University of Florida
Gainesville, Florida, U.S.A.

Faran Nouri
Applied Materials Inc.
Sunnyvale, California, U.S.A.

Wen-Chin Lee
TSMC Ltd.
Hsinchu, Taiwan

Wilman Tsai
Intel Corporation
Santa Clara, California, U.S.A.

Materials Research Society
Warrendale, Pennsylvania

CAMBRIDGE
UNIVERSITY PRESS

University Printing House, Cambridge CB2 8BS, United Kingdom

One Liberty Plaza, 20th Floor, New York, NY 10006, USA

477 Williamstown Road, Port Melbourne, VIC 3207, Australia

314-321, 3rd Floor, Plot 3, Splendor Forum, Jasola District Centre, New Delhi - 110025, India

79 Anson Road, #06-04/06, Singapore 079906

Cambridge University Press is part of the University of Cambridge.

It furthers the University's mission by disseminating knowledge in the pursuit of
education, learning and research at the highest international levels of excellence.

www.cambridge.org
Information on this title: www.cambridge.org/9781558998698

Materials Research Society
506 Keystone Drive, Warrendale, PA 15086
http://www.mrs.org

© Materials Research Society 2006

First published 2006
First paperback edition 2012

Single article reprints from this publication are available through
University Microfilms Inc., 300 North Zeeb Road, Ann Arbor, MI 48106

CODEN: MRSPDH

A catalogue record for this publication is available from the British Library

ISBN 978-1-558-99869-8 Hardback
ISBN 978-1-107-40883-8 Paperback

CONTENTS

*Invited Paper

PROCESS AND SUBSTRATE-INDUCED
STRAINED-Si DEVELOPMENT

POSTER SESSION

CHARACTERIZATION OF NEW MATERIALS
AND STRUCTURES

MODELING AND METROLOGY

*Invited Paper

PREFACE

For the past four decades, geometric scaling of silicon CMOS transistors has enabled not only an exponential increase in circuit integration density — Moore's Law — but also a corresponding enhancement in the transistor performance. Simple MOSFET geometric scaling has driven the industry to date. However, as the transistor gate lengths drop below 35 nm and the gate oxide thickness is reduced to 1 nm, physical limitations such as off-state leakage current and power density make geometric scaling an increasingly challenging task.

In order to continue CMOS device scaling, innovations in device structures and materials are required and the industry needs a new scaling vector. Starting at the 90 and 65 nm technology generation, *strained silicon* has emerged as one such innovation. Other device structures such as multi-gate FETs may be introduced to meet the scaling challenge.

Symposium D, "Transistor Scaling—Methods, Materials and Modeling," held on April 18-19 at the 2006 MRS Spring Meeting in San Francisco, California, had 54 oral and poster presentations, including eight invited papers from industry and academic leaders. The symposium brought together materials scientists, silicon technologists and TCAD researchers to share experimental results and physical models related to state-of-the-art MOSFETs, and to discuss the new and innovative approaches necessary to continue the transistor scaling. This volume contains expanded versions of many of these presentations in the areas of technology development, metrology, characterization and modeling.

We wish to thank all our invited speakers, our contributors and our participants for a lively and stimulating symposium. We also gratefully acknowledge the financial support from TSMC, Ltd.

Scott Thompson
Faran Nouri
Wen-Chin Lee
Wilman Tsai

September 2006

MATERIALS RESEARCH SOCIETY SYMPOSIUM PROCEEDINGS

MATERIALS RESEARCH SOCIETY SYMPOSIUM PROCEEDINGS

Volume 915— Nanostructured Materials and Hybrid Composites for Gas Sensors and Biomedical Applications, P.I. Gouma, D. Kubinski, E. Comini, V. Guidi, 2006, ISBN 978-1-55899-871-3

Volume 916— Solid-State Lighting Materials and Devices, F. Shahedipour-Sandvik, E.F. Schubert, B.K. Crone, H. Liu, Y-K. Su, 2006, ISBN 978-1-55899-872-1

Volume 917E— Gate Stack Scaling—Materials Selection, Role of Interfaces and Reliability Implications, R. Jammy, A. Shanware, V. Misra, Y. Tsunashima, S. De Gendt, 2006, ISBN 978-1-55899-874-8

Volume 918E— Chalcogenide Alloys for Reconfigurable Electronics, A.H. Edwards, P.C. Taylor, J. Maimon, A. Kolobov, 2006, ISBN 978-1-55899-875-6

Volume 919E— Negative Index Materials—From Microwave to Optical, S-Y. Wang, N.X. Fang, L. Thylen, M.S. Islam, 2006, ISBN 978-1-55899-876-4

Volume 920— Smart Nanotextiles, D. Diamond, X. Tao, G. Tröster, 2006, ISBN 978-1-55899-877-2

Volume 921E— Nanomanufacturing, F. Stellacci, J.W. Perry, G.S. Herman, R.N. Das, 2006, ISBN 978-1-55899-878-0

Volume 922E— Organic and Inorganic Nanotubes—From Molecular to Submicron Structures, K. Nielsch, O. Hayden, H. Ihara, D. Wang, 2006, ISBN 978-1-55899-879-9

Volume 923E— Structure and Dynamics of Charged Macromolecules at Solid-Liquid Interfaces, V.M. Prabhu, G. Fytas, A. Dhinojwala, 2006, ISBN 978-1-55899-880-2

Volume 924E— Mechanics of Nanoscale Materials and Devices, A. Misra, J.P. Sullivan, H. Huang, K. Lu, S. Asif, 2006, ISBN 978-1-55899-881-0

Volume 925E— Mechanotransduction and Engineered Cell-Surface Interactions, M.P. Sheetz, J.T. Groves, D. Discher, 2006, ISBN 978-1-55899-882-9

Volume 926E— Electrobiological Interfaces on Soft Substrates, J.P. Conde, B. Morrison III, S.P. Lacour, 2006, ISBN 978-1-55899-883-7

Volume 927E— Hydrogen Storage Materials, J.C.F. Wang, W. Tumas, Aline Rougier, M.J. Heben, E. Akiba, 2006, ISBN 978-1-55899-884-5

Volume 928E— Current and Future Trends of Functional Oxide Films, D. Kumar, V. Craciun, M. Alexe, K.K Singh, 2006, ISBN 978-1-55899-885-3

Volume 929— Materials in Extreme Environments, C. Mailhiot, P.B Saganti, D. Ila, 2006, ISBN 978-1-55899-886-1

Volume 930E— Materials Science of Water Purification, M.A. Shannon, D. Ginley, A.M. Weiss, 2006, ISBN 978-1-55899-887-X

Volume 931E— Education in Nanoscience and Engineering, R. Carpenter, S. Seal, N. Healy, N. Shinn, W. Braue, 2006, ISBN 978-1-55899-888-8

Volume 932— Scientific Basis for Nuclear Waste Management XXIX, Pierre Van Iseghem, 2006, ISBN 978-1-55899-889-6

Volume 933E— Science and Technology of Nonvolatile Memories, O. Auciello, J. Van Houdt, R. Carter, S. Hong, 2006

Volume 934E— Silicon-Based Microphotonics, M. Brongersma, D.S. Gardner, M. Lipson, J.H. Shin, 2006

Volume 935E— Materials Research for THz Applications, O. Mitrofanov, X-C. Zhang, R. Averitt, K. Hirakawa, A. Tredicucci, 2006

Volume 936E— Materials for Next-Generation Display Systems, F.B. McCormick, J. Kido, J. Rogers, S. Tokito, 2006

Volume 937E— Conjugated Organic Materials—Synthesis, Structure, Device and Applications, Z. Bao, A.B. Chwang, L. Loo, R.A. Segalman, 2006,

Volume 938E— Molecular-Scale Electronics, R. Shashidhar, J.G. Kushmerick, H.B. Weber, N.J. Tao, 2006

Volume 939E— Hybrid Organic/Inorganic/Metallic Electronic and Optical Devices, V. Bulovic, S. Coe-Sullivan, P. Peumans, 2006

Volume 940E— Semiconductor Nanowires—Fabrication, Physical Properties and Applications, M. Zacharias, W. Riess, P. Yang, Y. Xia, 2006

Volume 941E— Magnetic Thin Films, Heterostructures and Device Materials, T. Ambrose, W. Bailey, D. Keavney, Y.D. Park, 2006

Volume 942E— Colloidal Materials—Synthesis, Structure and Applications, N. Wagner, G.G. Fuller, J. Lewis, K. Higashitani, 2006

Volume 943E— Nanostructured Probes for Molecular Bio-Imaging, G. Bao, J. West, S. Nie, A. Tsourkas, 2006

Volume 944E— Molecular Motors, Nanomachines, and Engineered Bio-Hybrid Systems, S. Diez, B.R. Ratna, J.F. Stoddart, L. Turner, 2006

Volume 945E— Materials and Basic Research Needs for Solar Energy Conversion, P. Alivisatos, N.S. Lewis, A.J. Nozik, M. Wasielewski, 2006

Volume 946E— Recent Advances in Superconductivity, L. Civale, C. Cantoni, M. Feldman, X. Obradors, 2006

Prior Materials Research Society Symposium Proceedings available by contacting Materials Research Society

SOI, FDSOI, SGOI, GOI, Multi-Gate and Schottky SD Technologies

Mater. Res. Soc. Symp. Proc. Vol. 913 © 2006 Materials Research Society　0913-D01-01

Amorphization/Templated Recrystallization (ATR) Method for Hybrid Orientation Substrates

K. L. Saenger[1], J.P. de Souza[1], K.E. Fogel[1], J.A. Ott[1], A. Reznicek[1], C.Y. Sung[1], H. Yin[2], and D.K. Sadana[1]

[1]IBM Semiconductor Research and Development Center, T.J. Watson Research Center, P.O. Box 218, Yorktown Heights, New York, 10598
[2]IBM Semiconductor Research and Development Center, Microelectronics Division, Hopewell Junction, New York, 12533

ABSTRACT

Hybrid orientation substrates make it possible to have a CMOS technology in which nFETs are on (100) Si (the Si orientation in which electron mobility is the highest) and pFETs are on (110)-oriented Si (the Si orientation in which hole mobility is the highest). This talk will describe a new amorphization/templated recrystallization (ATR) method for fabricating bulk hybrid orientation substrates. In a preferred version of this method, a silicon layer with a (110) orientation is directly bonded to a Si base substrate with a (100) orientation. Si regions selected for an orientation change are amorphized by ion implantation and then recrystallized to the (100) orientation of the base substrate. After an overview of the ATR technique and its various implementations, we will describe some of the scientifically interesting materials and integration challenges encountered while reducing it to practice.

INTRODUCTION

The desire for improved CMOS device performance coupled with concerns about the limits of scaling is driving renewed interest in new materials (e.g., high-k gate dielectrics and metal gates) and structural variations of old materials (e.g., hybrid orientation substrates and strained Si) for field effect transistors (FETs). Hybrid orientation substrates, generally comprising a set of first semiconductor regions with a first crystal orientation and a second set of semiconductor regions with a second crystal orientation, are of interest because they allow a CMOS technology in which nFETs are on (100) Si (the Si orientation in which electron mobility is the highest) and pFETs are on (110)-oriented Si (the Si orientation in which hole mobility is the highest).

Hybrid orientation substrates have previously been fabricated with a bonding/epitaxial growth method in which regions of a bonded layer of Si having a first orientation are etched away and then replaced with epitaxially-grown Si having a second orientation matching that of the underlying Si substrate [1,2]. This paper describes an alternative amorphization/templated recrystallization (ATR) method for fabricating bulk hybrid orientation substrates [3], in which a silicon layer with a first orientation is directly bonded to a Si base substrate with a second orientation to form a direct-silicon-bonded (DSB) wafer. Regions in the bonded DSB layer selected for an orientation change are amorphized by ion implantation and then recrystallized to the orientation of the base substrate.

After illustrating the basic feasibility of the ATR method and showing examples of some simple integration schemes, we will describe a selection of the scientifically interesting materials and integration challenges encountered while reducing the method to practice. Particular attention will be directed towards problems and progress in three areas: initial substrate preparation; the quality of the changed-orientation ATR'd material; and "edge defects" associated with patterned ATR. Experimental details not included in the next section will be introduced in the individual topic sections as needed.

EXPERIMENTAL DETAILS

Amorphization for ATR was effected by ion implantation of Si^+ or Ge^+ ions, with the energy and dose selected to insure complete amorphization from the substrate's top surface to a depth below the DSB interface. All implants utilized a $7°$ tilt (to minimize channeling effects) and 10 °C substrate cooling. The thicknesses of the amorphized layer and the highly damaged crystalline layer just below it (located at the end-of-range (e-o-r) of the implanted ions) were determined by cross-section scanning electron microscopy (SEM) on Cr-coated samples Secco-etched after cleaving.

Templated recrystallization was typically performed at temperatures between 650 and 900 °C for times ranging from 1 to 10 min, though additional defect removal anneals (at 1300 to 1325 °C for 3 to 5 hours) were also performed after deposition of protective oxide capping layers. Plan view SEM (after Secco etching) was used to evaluate areal defect densities and to identify defective regions at the edges of patterned ATR'd regions. Cross section transmission electron microscopy (XTEM) was used to examine the samples at various stages of processing, with microdiffraction used to verify the expected Si orientation.

ATR: BASIC METHOD AND INTEGRATION SCHEMES

The basic ATR technique is shown schematically in Fig. 1. The starting DSB substrate (Fig. 1a) comprises a handle Si wafer of a first orientation (j'k'l') and a DSB Si device layer of a second orientation (jkl). The DSB layer is amorphized by ion implantation (I/I) to create an amorphous Si (a-Si) layer

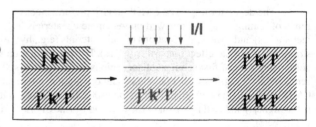

Figure 1. A schematic of the basic ATR technique.

extending from the substrate surface down to a depth below the bonded interface (Fig. 1b), and then recrystallized to the orientation of the underlying handle wafer (Fig. 1c). For convenience, we will designate wafers with (110) DSB layers on (100) substrates as DSB-A, and wafers with (100) DSB layers on (110) substrates DSB-B.

Fabrication of hybrid orientation substrates requires Si orientation changes in selected regions only (i.e., patterned ATR vs. the maskless ATR of Fig. 1). Fig. 2 shows two versions of patterned ATR. In the first, the ATR is performed without shallow trench isolation (STI). In the second, the ATR is performed after STI formation. A priori, the ATR-with-STI version might appear preferable because the trenches provide alignment marks (a necessity, since the differently-oriented Si regions are optically indistinguishable) and eliminate the possibility of lateral templating.

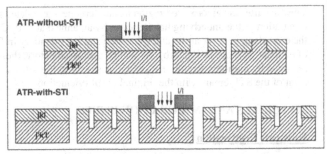

Figure 2. Two versions of patterned ATR.

INITIAL REDUCTION TO PRACTICE

We have performed blanket ATR to achieve both (110)-to-(100) and (100)-to-(110) conversions in DSB-A and DSB-B wafers respectively. The DSB-A conversion is illustrated in the Fig. 3 with XTEM images of a 50 nm-thick (110)-oriented DSB overlayer on a (100)-oriented Si substrate before (a) and after (b) a blanket ATR process that changed the DSB layer orientation to (100). In this case the amorphization implant was 2.0×10^{15} /cm^2 220 keV Ge$^+$ and the recrystallization anneal was at 900 °C for 1 min. No sign of the bonded interface remains after ATR. As will be discussed in more detail later, the ATR'd Si is relatively free of defects, though a band of dislocation loops remains at the position of the end-of-range (e-o-r) damage layer.

The XTEM images of Fig. 4 show the results for selected-area ATR on substrates patterned with oxide-filled trenches using the ATR-with-STI integration scheme of Fig. 2. Fig. 4(a) shows the substrate after a region of the 200 nm-thick (110) DSB layer to the left of the STI has been amorphized by an implant of 2.5 x 10^{15} /cm^2 120 keV Si$^+$, and Fig. 4(b) shows the

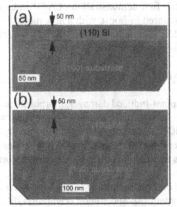

Figure 3. XTEM images showing the (110)-to-(100) orientation change of a 50-nm-thick Si layer on a (100)-oriented Si substrate: before ATR, (a); after ATR, (b).

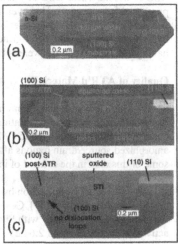

Figure 4. XTEM images showing the selected area (110)-to-(100) orientation change of a 200-nm-thick DSB layer on a (100)-oriented Si substrate.

substrate after the amorphized region has undergone a templated recrystallization to the (100) orientation of the underlying handle wafer by an anneal at 650 °C for 5 min. Clearly visible are the e-o-r damage layer at the amorphous/crystalline boundary in Fig. 4(a), and the resulting band of dislocation loops in Fig. 4(b). As will be discussed below, these damage regions may be removed by high-temperature annealing, as shown in Fig. 4(c). As expected, DSB regions to the right of the STI remain with the original (110) orientation.

MATERIALS AND INTEGRATION ISSUES

Substrate Preparation

Si wafer bonding techniques are generally either hydrophobic (in which H-terminated surfaces are bonded) or hydrophilic (in which OH-terminated, oxide-like surfaces are bonded). Hydrophilic bonding is now a mature technology, in large part due to industry demand for silicon-on-insulator (SOI) wafers. These SOI wafers typically comprise a Si layer disposed on a buried oxide (box) layer, with the box layer thickness being determined by the amount of surface oxide on the two starting wafers.

The desired box layer thickness for DSB wafers subjected to orientation-changing ATR is zero. In principle, such wafers would best be fabricated by hydrophobic bonding. However, hydrophobic bonding techniques are both more difficult (in part due to the propensity of H-terminated surfaces to attract particulates that interfere with bonding) and less commonly practiced (since they are not needed for SOI wafers).

For reasons that will become apparent in the next section, most of our work has been with DSB-A wafers. These DSB-A wafers, with DSB layers 50 to 200 nm in thickness, were typically purchased from vendors. However, our first demonstration of ATR-induced orientation change was performed on a DSB-B wafer prepared in-house with a new method in which standard hydrophilic bonding and wafer grinding techniques were used to create mixed orientation wafers having thin box layers that were subsequently removed *in situ*.

Quality of ATR'd Material

Much is known about ion implant-induced damage/amorphization [4,5] and templated recrystallization in single-orientation bulk Si [6,7], in part because ATR without an orientation change is routinely employed in CMOS processing to recrystallize amorphized or pre-amorphized source/drain regions after dopant implantation. In this section we will briefly review some of this data in the context of its relation to the quality of changed-orientation Si.

Rates of recrystallization by solid phase epitaxy (SPE) have been studied as a function of temperature [8], Si dopant [7], Ge content [8], and silicon orientation [6]. Recrystallization rates for undoped Si amorphized with Si are quite fast, about 7.3 nm/s for (100) Si at 650 °C, with an activation energy of about 2.64 eV. Recrystallization rates are orientation-dependent, with recrystallization fastest in (100)-oriented Si and slower by factors of about 2.5 for (110)-oriented Si and 10 for (111)-oriented Si. Defect densities in Si recrystallized by SPE also depend on Si crystal orientation, being lowest to highest in the order (100) < (110) < (111).

SPE rates of I/I-amorphized Si are also dopant dependent. For (100)-Si doped at ~2 x 10^{20}/cm^3 (0.4 at %), the rates are faster for B (x20), P (x6), and As (x5); the same for Ge; and slower for Ar (factor of 90), O and N (factor of ~9), and C (factor of 1.6); where all rates are referenced to Si amorphized by self-implantation with Si$^+$.

In our own defect studies we mostly used (110) or (100) bulk wafers to avoid introducing the bonded interface as a variable. Two types of defects are the most prevalent: e-o-r damage loops (initiated by the interstitials left where the implanted ions stop), and "threads" which show up as pits in plan view SEMs after Secco etching. For the same implant and annealing conditions, pit density was always at least 10x higher in (110) Si than (100) Si, a clear reason to prefer DSB-A substrates over DSB-B. In general, surface defects are worse when the initial amorphous/crystalline Si interface is rough, so better SPE would be expected with substrate cooling during implantation, heavier ions (e.g., Ge$^+$ vs. Si$^+$) and a thinner amorphized layer (though this can only be taken so far, since the minimum amorphization depth is constrained by the thickness of the DSB layer).

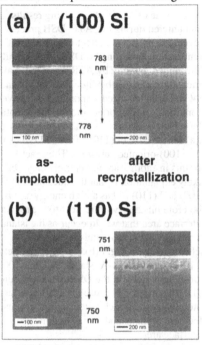

Fig. 5 compares the cross section SEM images for (100) and (110) Si amorphized by a dual implant of 1.0 x 10^{15}/cm^2 50 keV Si$^+$ followed by 4.0 x 10^{15}/cm^2 220 keV Si^{++} (equivalent to 440 keV Si$^+$). This implant produced an amorphous layer about 750-780 nm in thickness bounded below by a crystalline damage layer. After recrystallization (by annealing at 650 °C for 5 min), a layer of e-o-r loops near the position of the initial amorphous Si/crystalline Si interface is clearly visible. Also evident is the strikingly more defective recrystallization of the (110) material. However, as discussed above, these are "worst case" ATR conditions, since the amorphizing species is a light ion (Si$^+$) and the amorphized layer is quite thick.

Figure 5. A comparison of blanket ATR in (100) and (110) Si showing the amorphized layer in the as-implanted samples before and after a recrystallization anneal.

It was found that pit/thread density can be dramatically lowered and the e-o-r damage loops can be made to completely disappear by annealing at high temperature (1300 - 1325 °C) for several hours in an inert ambient. Normally such processing temperatures are incompatible with CMOS processing due to concerns about dopant diffusion, but this is not an issue at these early stages of substrate preparation. Our high temperature defect removal process is shown schematically in Fig. 6 and was used to remove e-o-r loops from ATR'd Si recrystallizing to a (100) orientation as shown in the XTEM sample of Fig. 4(c). This same process was also effective in reducing the defectivity of ATR'd Si recrystallizing to a (110) orientation, as can be seen from the SEM image of Fig. 7 for a sample amorphized and recrystallized with the conditions of the samples shown in Fig. 5. However, there is no doubt that the true measure of the quality of the ATR'd material can only come from device data.

One of the most surprising results encountered during our ATR-DSB process development was the effect of this high temperature annealing on DSB-layer islands surrounded laterally by ATR'd material having the orientation of the substrate. The DSB islands were not stable and showed a spontaneous conversion from their original orientation to the orientation of the substrate. This is shown schematically in Fig. 8(a) for the case of (110)-oriented DSB-A islands in a (100)-oriented substrate. The exact reasons for this instability are not entirely clear, but a likely explanation is that the interface between the (100) and (110) Si has a high energy and is therefore unstable with respect to the reduction in interface area that would occur as the island shrinks in size. Fortunately, as shown schematically in Fig. 8(b), the interface between the (110)-oriented DSB-A islands and the (100)-oriented substrate is stable if the DSB island is bounded by STI. This may be because the DSB interface is pinned laterally by the trench oxide, and/or because a decrease in the island size would not reduce the interface area.

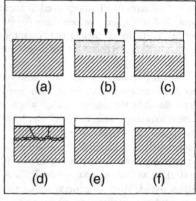

Figure 6. A schematic of the defect removal process, starting with the initial substrate, (a); continuing with implantation to make an amorphous layer, (b); deposition of a protective cap, (c); recrystallization with defects, (d); high temperature annealing to remove defects, (e); and cap removal, (f).

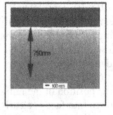

Figure 7. The (110) sample of Fig. 5(b) after the high temperature defect removal process shown in Fig. 6.

Figure 8. A schematic showing the stability (or instability) of (110)-oriented DSB-A islands in a (100)-oriented substrate to high temperature annealing: islands not bounded by STI, (a); islands bounded by STI, (b).

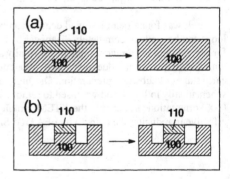

Defects with Patterned ATR

In this section we will discuss defects peculiar to patterned ATR: the trench-edge defects seen with ATR-with-STI, and the border region morphology seen with ATR-without-STI.

We begin with the trench-edge defects seen with ATR-with-STI. While such defects have been discussed previously [9] and are frequently present at the edges of source/drain regions that have undergone amorphizing implants, they present particular concern for ATR-DSB processes, since the FET gates crossing over to isolation regions cannot avoid passing over the defective Si edges. The problem is shown schematically in Fig. 9 for the case of (100) Si, and illustrated with examples from our own work in Fig. 10a (a higher magnification XTEM of the sample of Fig. 4b) and in Fig. 10b (SEM of a similar DSB sample amorphized with slightly different implant conditions).

Figure 9. A schematic of trench-edge defects using ATR with DSB-A and recrystallization to the substrate (100) orientation.

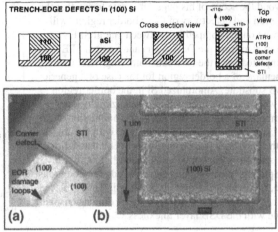

Fig. 10. XTEM image of a trench-edge defect in the sample of Fig. 4(b), (a); a plan view SEM image of a similar sample given a slightly different amorphization implant, (b).

Burbure and Jones [9] attribute these trench edge defects to poor epitaxy on (111) planes during a process sequence shown schematically in Fig. 11. Oxide-filled trenches are first formed around the boundaries of a single-crystal Si region, followed by amorphization of these trench-bounded Si regions by a maskless ion implantation. At an intermediate stage of annealing, recrystallization at the feature edges stops on (111) planes originating from the three-phase intersection of the initial amorphous/crystalline interface and the oxide-filled trench. Recrystallization of the remaining amorphous edge regions must thus templated from (111) planes (the crystal orientation on which SPE defects are the worst), producing the defective final structure shown. Since the (111) planes intersect all four oxide trench edges at a fixed 35.3° angle, defects will be on all four trench edges and the lateral extent of the trench-edge defects will scale with the amorphization depth, as observed previously [9].

Figure 11. Process sequence for trench-edge defect formation, as described in Ref. 9.

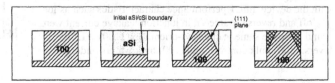

Our own observations of trench-edge defects in (110) Si, shown schematically in Fig. 12, are consistent with this model. We see defective edge regions on two sides instead of four, and, for the same amorphization depth, wider defects on the bad edges than those in (100), because the (111) planes intersect only two trench edges and do so at a wider angle (54.7° instead of 35.3°).

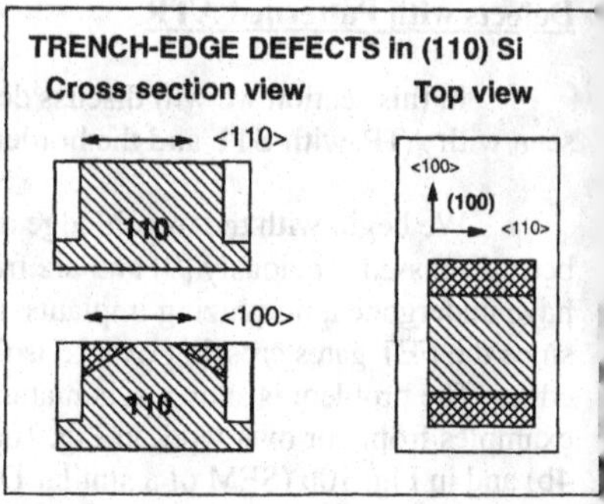

Process flows using ATR-without-STI avoid the issue of trench-edge defects, but introduce the possibility of lateral templating at the patterned ATR edges. While a detailed discussion of this topic is beyond the scope of his paper, it is safe to say that the location and defectivity of these border regions will depend on the competition between lateral SPE (templated from the DSB layer) and vertical SPE (templated from the base substrate). Fig. 13 shows a schematic of a border region for the case of a generic DSB substrate. Amorphized Si in the border region between the dark lines will recrystallize into some combination of the DSB layer's (jkl) orientation and the base substrate's (j'k'l') orientation, with the angles of the dark lines fixed by the growth characteristics of the specific crystal orientations contributing to the templating. Whatever this defective edge region looks like, however, it is expected that it will (i) be replaced by an STI trench, and (ii) have lateral dimensions that scale with the DSB layer thickness.

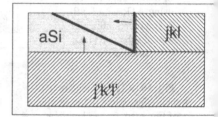

Figure 12. A schematic of trench-edge defects in (110) Si.

Figure 13. A schematic of a border region during patterned ATR. The amorphized Si within the dark lines will recrystallize to some combination of the (jkl) or (j'k'l') orientations.

DEVICE RESULTS

In this section we summarize the recent device measurements of Ref. 10 for bulk CMOS FETs with Lpoly = 45 nm fabricated on hybrid orientation substrates made by DSB-ATR from DSB-A wafers. In this case, the nFETs are in changed-orientation (100) Si, and the pFETs are in the original (110) DSB layer, as shown schematically in Fig. 14. pFET off current (Ion) vs. saturation current (Ion) curves were found to be the same in both the original (110) DSB layers and (110) bulk controls, implying that the DSB layer is not degraded by processing, and that the bonded interface does not interfere with the device. nFET performance characteristics such as Ion vs. Ioff and reverse source/drain junction leakage current were found to be the same in changed-orientation (100) ATR'd layers, (100) bulk controls, and (100) bulk wafers undergoing

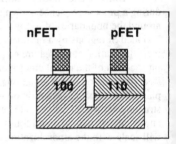

Figure 14. The preferred arrangement of nFETs and pFETs on a hybrid orientation substrate.

ATR with no orientation change (i.e., amorphization followed by recrystallization back into the substrate's original (100) orientation), implying no degradation from defects generated by the ATR process. Ring oscillators fabricated on these same wafers showed improvements of more than 20% compared with the same devices on the (100) control wafers.

SUMMARY

We have shown that the crystal orientation of single-crystal silicon layers may be changed in selected areas from one orientation to another by an amorphization/templated recrystallization (ATR) process on direct-silicon-bonded (DSB) wafers, and described practical ways in which the ATR-DSB method might be use to form hybrid orientation substrates. The materials and integration challenges encountered while reducing the method to practice have led to some interesting and unexpected results, including some which are still under investigation. Findings of particular interest include (i) observations of oxide removal from a buried interface (with applications to hydrophilic methods of forming DSB wafers), (ii) the use of high-temperature annealing to remove e-o-r damage (as well as embedded islands of differently-oriented Si), and (iii) the differences in trench edge defects between (100) Si (the case of DSB-A) and (110) Si (the case of DSB-B).

ACKNOWLEDGMENTS

The authors wish to thank the personnel of the Yorktown and East Fishkill fabrication facilities for help with substrate preparation and processing.

REFERENCES

1. M. Yang, M. Ieong, L. Shi, K. Chan, V. Chan, A. Chou, E. Gusev, K. Jenkins, D. Boyd, Y. Ninomiya, D. Pendleton, Y. Surpris, D. Heenan, J. Ott, K. Guarini, C. D'Emic, P. Saunders, M. Cobb, P. Mooney, B. To, N. Rovedo, J. Benedict, and H. Ng, International Electron Devices Meeting Tech. Dig. 453 (2003).
2. S. Yoshikawa and A. Sudo, U.S. Patent No. 5,384,473 (24 January 1995).
3. K.L. Saenger, J.P. de Souza, K.E. Fogel, J.A. Ott, A. Reznicek, C.Y. Sung, D.K. Sadana, and H. Yin, Appl. Phys. Lett. 87 221911 (2005).
4. J. P. de Souza and D.K. Sadana, in *Handbook on Semiconductors: Materials, Properties and Preparation,* edited by S. Mahajan (Elsevier North-Holland, 1994), Vol. 3b, p. 2033.
5. S. Milita and M. Servidori, J. Appl. Phys 79 8278 (1996).
6. L. Csepregi, E.F. Kennedy, J.W. Mayer, and T.W. Sigmon, J. Appl. Phys. 49 3906 (1978).
7. L. Csepregi, E.F. Kennedy, T.J. Gallagher, J.W. Mayer, and T.W. Sigmon, J. Appl. Phys. 48 4234 (1977).
8. T.E. Haynes, M.J. Antonell, C. Archie Lee, and K.S. Jones, Phys. Rev. B 51 7762 (1995).
9. N. Burbure and K.S. Jones, Mat. Res. Soc. Symp. Proc. 810 C4.19.1 (2004).
10. C.-Y. Sung, H. Yin, H.Y. Ng, K.L. Saenger, V. Chan, S.W. Crowder, J. Li, J.A. Ott, R. Bendernagel, J.J. Kempisty, V. Ku, H.K. Lee, Z. Luo, A. Madan, R.T. Mo, P.Y. Nguyen, G. Pfeiffer, M. Raccioppo, N. Rovedo, D. Sadana, J.P. de Souza, R. Zhang, Z. Ren. and C.H. Wann, 2005 IEEE International Electron Devices Meeting, Paper 10.3 (2005).

Mater. Res. Soc. Symp. Proc. Vol. 913 © 2006 Materials Research Society   0913-D01-02

Systematic Characterization of Pseudomorphic (110) Intrinsic SiGe Epitaxial Films for Hybrid Orientation Technology with Embedded SiGe Source/Drain

Qiqing (Christine) Ouyang[1], Anita Madan[2], Nancy Klymko[2], Jinghong Li[2], Richard Murphy[2], Horatio Wildman[2], Robert Davis[2], Conal Murray[1], Judson Holt[2], Siddhartha Panda[2], Meikei Ieong[2], and Chun-Yung Sung[2]

[1]T J Watson Research Center, S&TG, IBM, Route 134, Yorktown Heights, NY, 10598
[2]S&TG, IBM, 2070 Route 52, Hopewell Junction, NY, 12533

ABSTRACT

PFETs with embedded SiGe source/drain on HOT substrates have shown significant performance improvement compared to PFETs with embedded SiGe on (100) SOI substrates. In this paper, we report a systematic material characterization on the epitaxial SiGe films, both blanket and patterned, on (110) and (100) substrates, using an array of methods such as XRD AES, UV Raman, AFM, and TEM, and corresponding PFETs performance data.

INTRODUCTION

To exploit the higher mobility of holes on (110) and electrons on (100) substrates, two types of techniques have been developed in order to monolithically integrate different crystal orientations -- one is the hybrid orientation technology (HOT) [1]; and the other is the mixed orientation formed by direct silicon bonding (DSB) and followed by a solid phase epitaxy (SPE) [2]. Meanwhile, various techniques have been used to introduce uniaxial stress in MOSFET channel areas and improve the performance of the conventional devices. In particular, SiGe embedded in the source/drain (S/D) on HOT has been demonstrated to further enhance the pFET performance [3]. In this paper, we report a systematic study on the intrinsic pseudomorphic SiGe epi films on (110) Si substrate and explore the opportunities of fabricating strained channels on HOT substrates.

EXPERIMENTAL DETAILS

Commercially available semiconductor equipment for rapid thermal chemical vapor deposition (RTCVD) was used to deposit $Si_{(1-x)}Ge_x$ (where 'x' is the Ge composition) thin films with basic germane and DCS chemistry. Blanket films of nominal 10% and 15% Ge and thicknesses 49-90 nm were deposited on 300 mm (100) and (110) silicon wafers at a temperature of 750^0C and well characterized before deposition on patterned device wafers. Epitaxial growth rate on (110) substrates was 30% lower than that of (100), while it was 30% higher on patterned wafers compared to blanket wafers because of loading effects.

These SiGe films were characterized using X-ray Diffraction (XRD), Auger and Raman to correlate different techniques for measuring Ge composition and the amount of strain. A Bede X-ray diffractometer using monochromatic CuKα (wavelength λ = 1.5406Å) radiation was used for rocking curve measurements around the Si (110) reflection. Auger Electron Spectroscopy (AES) data was collected on either a ULVAC-PHI SmartTool or Model 650 scanning Auger microprobe, using 20 nA (ST) or 100 nA (650) at 10 kV. Alternate sputtering with a 2 kV Ar+ beam was used for depth profiling. For quantification, undoped blanket SiGe films, analyzed by Rutherford backscattering (RBS), were used to obtain average elemental sensitivity factors for each tool, correcting for instrumental response as well as preferential sputtering. Raman measurements were made using a HoribaJY LabRam 800 spectrometer system with microprobe capability. The excitation wavelength was 325 nm allowing for approximately 15 nm sampling depth into the SiGe. Using a 0.9NA UV objective (Leica), the lateral spatial resolution was approximately 0.7 μm. Atomic Force measurements (AFM) were performed on a Veeco Instruments Nanoscope D-5000 AFM, using an etched silicon tip with radius of 5-10 nm. The roughness calculations were taken from a 10 μm x 10 μm area. The Rmax represented the maximum peak to valley value within the area analyzed. The RMS value was calculated as the root mean square average of height deviations taken from the mean data plane. Transmission Electron Microscopy (TEM) was carried out on a JEOL 2010 field emission microscope, while the samples were prepared using industry standard focused ion beam technique.

DISCUSSION

Figure 1(a) shows the sample rocking curve scans around the (110) reflection for SiGe of different thicknesses. Clearly defined periodic fringes are seen in both symmetric scans suggesting that the film growth is pseudomorphic. The interference fringe spacing $\Delta\theta_p$ and the peak spacing $\delta\theta$ between the Si substrate and the SiGe peak were used to determine the film thickness and the perpendicular strain [4]. The fringe spacing $\Delta\theta_p$ was measured to be 190" and 360" and the thickness t of the film was calculated to be 49 nm and 90 nm using equation (1)

$$t = \lambda / \Delta\theta_p \cos\theta \qquad (1)$$

where θ is the angle of incidence and diffraction of the X-ray beam relative to the reflecting plane. The out-of-plane SiGe and Si lattice parameters, c_{SiGe} and a_{Si}, respectively, can be approximated by using the differential form of Bragg's law:

$$(c_{SiGe} - a_{Si})/a_{Si} \approx - (\theta_{SiGe} - \theta_{Si}) \cot(\theta_{Si}) \qquad (2)$$

Assuming a fully strained SiGe film on an unstrained Si substrate, c_{SiGe} can be related to the unstrained SiGe lattice parameter, a_{SiGe}, using linear elasticity theory:

$$c_{SiGe} = a_{SiGe}(1+\varepsilon_{zz}) = a_{SiGe}(1 + f \Delta\varepsilon) \qquad (3)$$

where ε_{zz} represents the out-of-plane SiGe strain, $\Delta\varepsilon$ the SiGe in-plane biaxial strain, and f is a parameter determined from the SiGe elastic constants. For an isotropic material, $f = -2\nu/(1-\nu)$ where ν is the Poisson's ratio of the SiGe film. However, f can be calculated for different orientations using a Vegard's law approximation for the components of the anisotropic elastic compliance tensor [5]. For a Ge fraction of 15%, f is calculated to be -0.501, resulting in an effective Poisson's ratio of 0.200 for a (110)-oriented SiGe film. These values are in contrast to those for a (100) oriented SiGe film, $f = -0.766$ and $\nu = 0.277$, possessing the same Ge fraction.

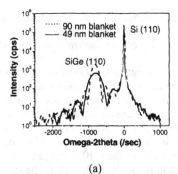
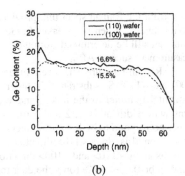

(a)   (b)

Figure 1(a): (110) rocking curve scans around the Si (110) reflection for two blanket SiGe films on (110) Si wafers; (b) Ge depth profile on two patterned device wafers measured using Auger electron spectroscopy.

Because $\Delta\varepsilon$ can be calculated from the lattice mismatch between the SiGe film and the underlying Si: $\Delta\varepsilon = (a_{Si} - a_{SiGe}) / a_{SiGe}$, Eqn's 2 and 3 can be simplified to form:

$$(\theta_{Si} - \theta_{SiGe}) \cot(\theta_{Si}) = (1 - f)(a_{SiGe} - a_{Si}) / a_{Si} \qquad (4)$$

From the measured diffraction peak positions, we calculate a biaxial strain $\Delta\varepsilon$ of -0.61% and, from the unstrained SiGe lattice parameter, a corresponding Ge fraction of 15.5% according to Dismukes et al. [6].

The Ge profiles were also measured with AES on both blanket and patterned epi films. The results are very close to those of XRD. For patterned wafers, AES was done on 100 μm structures. As shown in Figure 1(b), the Ge concentrations of 15.5% on the (100) wafer and 16.6% on the (110) wafer were found to be fairly uniform throughout the films. The data were then used to extract the strain in the Raman measurements.

The Raman phonon spectrum consists of three first-order lines corresponding to the nearest neighbor Si-Si, Si-Ge, and Ge-Ge atomic vibrations. The Ge-Ge line is very weak and not shown in Figure 2(b). The peak positions and relative intensities of the phonon lines depend on the Ge fraction, and the peak positions also depend on the strain, according to Eqn's 5a–c

$$\omega_{SiSi} = 521.0 - 68\,x + \Delta_{Si}\Sigma \qquad (5a)$$
$$\omega_{SiGe} = 400.5 + 14.2\,x + \Delta_{SiGe}\Sigma \qquad (5b)$$
$$\omega_{GeGe} = 282.5 + 16\,x + \Delta_{GeGe}\Sigma \qquad (5c)$$

where ω_{SiSi}, ω_{SiGe}, and ω_{GeGe} are the measured phonon peak positions; x is the Ge concentration; Σ is the normalized strain, i.e. ε / 0.0417; and 0.0417 is the mismatch strain between pure Ge and Si, making $\Sigma = 1$ for pure Ge grown epitaxially on (100) Si wafers [7].

The coefficients Δ for determining both 'x' and strain have been reported in the literature [7-9] and also using our own reference samples of epitaxial SiGe films on (100) silicon in the range $0.15 < x < 0.35$, which were independently characterized by RBS and XRD for Ge concentration (x) and strain (ε). Using Equations (5) it is possible to determine both 'x' and strain from the Raman spectrum, provided there is enough intensity in both the Si-Si and Si-Ge phonon lines. When the weaker Si-Ge line cannot be measured with adequate precision, the strain can still be determined from the Si-Si line, using the Ge composition from an independent measurement such as AES.

Raman measurements were made on both blanket and patterned SiGe films deposited on (100) and (110) Si substrates. In order to study the potential pattern-dependent effects, long horizontal (parallel to the notch) and vertical (perpendicular to the notch) stripes of SiGe epi films with width of 1.25, 2 and 5 µm, shown in Figure 2(a), were measured. While these epi stripes correspond to both <110> directions on (100) wafers, they are different on (110) wafers – the horizontal is along <110> direction, and the vertical is along <100> direction. The strain in the films along <110> and <100> directions with various width on the (110) patterned wafer are found to be 0.63~0.67% (only the data for 1.25µm epi strip is shown in Fig. 2(b)). The strain is the same as the strain on the (100) patterned sample, corresponding to fully strained $Si_{0.84}Ge_{0.16}$ films. Furthermore, the strain is in good agreement with those extracted from Raman and XRD on blanket films with the same alloy composition. All above is shown in Figure 2(b).

The surface roughness of the epi films was measured by AFM. Figure 3 shows the example AFM images of two blanket SiGe films grown on (110) and (100) Si substrates. RMS and Rmax comparison of blanket SiGe films with 15% are shown in the Table I. As the received Si wafers (out of box and no SiGe epi) had a RMS roughness of 0.061 and 0.095 nm for the (100) and (110) orientation, respectively, the (110) epi films had a 7x higher RMS roughness compared to the (100) epi films with the same deposition conditions.

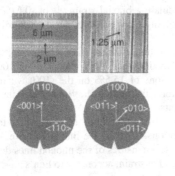

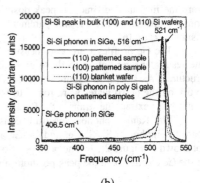

(a)   (b)

Figure 2(a) Locations of Raman microprobe measurements for the SiGe epi stripe regions on a patterned (110) wafers, and illustration of channel directions on (110) and (100) wafers; (b) Raman spectra for the SiGe epi films on the (110) and (100) patterned (epi width of 1.25µm) and blanket wafers.

Table I: AFM RMS and Rmax for blanket SiGe films on (110) and (100) Si substrates

Subs. Orientation	(110)	(110)	(110)	(100)	(100)	(100)
SiGe thickness (nm)	50	90	none	50	90	none
RMS(nm)	0.518	0.679	0.095	0.083	0.090	0.061
Rmax(nm)	10.264	5.135	0.961	0.773	0.872	0.690

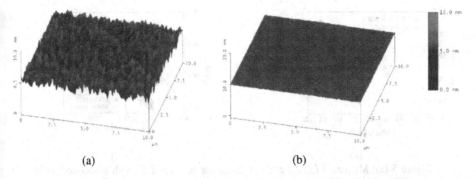

(a)   (b)

Figure 3: AFM images of 90 nm thick SiGe films grown on (a): (110); and (b): (100) Si substrates.

Figure 4(a) shows a weak beam dark field plan view TEM image of a blanket SiGe thin film grown on (110) Si substrate. Ge content is 15% and the thickness is 49 nm. No visible defects and no misfit dislocations were observed with a resolution of a few nanometers, indicating that the quality of the SiGe is very good, and that the film is fully strained. The same was true for a 90 nm thick film of similar Ge composition.

Devices with eSiGe S/D were fabricated using the 15% Ge epi process. Figure 4(b) shows a two-beam bright-field cross-sectional TEM image of a pFET device structure with embedded SiGe in the source/drain region on (110) Si. The epi has good quality and is fully strained. Only very low countable number of stacking faults was observed at the top corner of the SiGe source/drain.

Figure 5(a) and (b) show I_{dlin}-I_{off} and I_{on}-I_{off} for the 50 nm pFETs with <110>-oriented channels and eSiGe S/D on HOT and control substrates. I_{dlin} (I_d at $V_{gs} = 1$ V and $V_{ds} = 50$ mV) and I_{on} (I_d at $V_{gs} = V_{ds} = 1$ V) were improved by 57% and 30%, respectively, while threshold voltage and subthreshold swing were well matched between the devices on HOT and control substrates.

(a)   (b)

Figure 4 (a): PTEM image of a blanket SiGe film on a 110 Si substrate; (b): XTEM image of a pFET structure with embedded SiGe S/D on a HOT substrate.

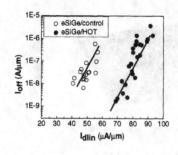 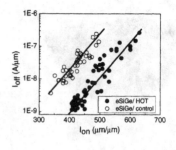

(a)   (b)

Figure 5 (a): Measured I_{dlin}-I_{off} and (b) I_{on}-I_{off} for 50nm pFETs with embedded SiGe S/D on HOT and (100) SOI substrates. Channels are <110> oriented.

CONCLUSIONS

Systematic materials characterization has been performed to study both blanket and patterned SiGe films epitaxially grown on (110) and (100) Si substrates. Page: 6 Consistent results have been found in determining the Ge composition using XRD and Auger, and in determining the strain using Raman and XRD. High quality and fully strained films have been achieved. Device performance enhancement has been demonstrated with eSiGe source/drain on HOT substrates.

REFERENCES

[1] M. Yang, M. Ieong, L. Shi, K. Chan, V. Chan, A. Chou, E. Gusev, K. Jenkins, D. Boyd, Y. Ninomiya, D. Pendleton, Y. Surpris, D. Heenan, J. Ott, K. Guarini, C. D'Emic, M. Cobb, P. Mooney, B. To, N. Rovedo, J. Benedict, R. Mo and H. Ng, IEDM Tech. Dig., 453 (2003).

[2] C.-Y. Sung, H. Yin, H. Y. Ng, K. L. Saenger, V. Chan, S. W. Crowder, J. Li, J. A. Ott, R. Bendernagel, J. J. Kempisty, V. Ku, H.K. Lee, Z. Luo, A. Madan, R.T. Mo, P.Y. Nguyen, G. Pfeiffer, M. Raccioppo, N. Rovedo, D. Sadana, J.P. de Souza, R. Zhang, Z. Ren and C.H. Wann, IEDM Tech. Dig., 235 (2005).

[3] Q. Ouyang, M. Yang, J. Holt, S. Panda, H. Chen, H. Utomo, M. Fischetti, N. Rovedo, J. Li, N. Klymko, H. Wildman, T. Kanarsky, G. Costrini, D. Fried, A. Bryant, J. A. Ott, M. Ieong and C.-Y. Sung, VLSI Symp. on Technologies, 28, (2005).

[4] D.K. Bowen and B. Tanner *High Resolution X-ray Diffraction and Topography*, 1st ed. (Taylor and Francis Ltd. London, 1998) p. 55-64.

[5] W.A. Brantley, J. Appl. Phys., 44(1), 534 (1973).

[6] J.P. Dismukes, L. Ekstrom, and R.J. Paff, J. Phys. Chem., 68(10), 3021 (1964).

[7] J.C. Tsang, P.M. Mooney, F. Dacol and J. Chu, J. Appl. Phys. 75(12), 8098 (1994).

[8] A. Tiberj, V. Paillard, C. Aulnette, N. Daval, K.K. Bourdelle, M. Moreau, M. Kennard, and I. Cayrefourcq, in *High-Mobility Group-1V Materials and Devices*, edited by M. Caymax. K. Rim, S. Zaima, E. Kasper and P.F.P. Fichtner (Mater. Res. Soc. Symp. Proc. 809, 2004) pp. 97-102.

[9] K. Brunner, in *Properties of Silicon Germanium and SiGe:Carbon*, edited by E. Kasper and K. Lyutovich, (INSPEC, The Institution of Electrical Engineers, London, UK, 2000) p.115.

Mater. Res. Soc. Symp. Proc. Vol. 913 © 2006 Materials Research Society   0913-D01-04

Schottky Source/Drain Transistor on Thin SiGe on Insulator Integrated with HfO2/TaN Gate Stack

Fei GAO[1,2], S.J. Lee[1], Rui Li[1], S. Balakumar[2], Chih-Hang Tung[2], Dong-Zhi Chi[3], and Dim-Lee Kwong[2]

[1]Department of ECE, Silicon Nano Device Lab, National University of Singapore, Block E4A #02-04 Engineering Drive 3, Singapore, 117576, Singapore
[2]Institute of Microelectronics Engineering, Singapore, Singapore, 117685, Singapore
[3]Institute of Materials Research Engineering, Singapore, 117602, Singapore

ABSTRACT

We report thin SGOI (Silicon Germanium on Insulator) with 65% Ge concentration p-MOSFET (Metal-Oxide-Semiconductor-Field-Effect-Transistor) using Ni-germanosilicide Schottky S/D (source/drain) and HfO_2/TaN gate stack integrated with conventional self-aligned top gate process. Unlike high temperature S/D activation needed for conventional transistor, low Ni-germanosilicide S/D formation temperature contributes to the excellent capacitance-voltage characteristic, low gate leakage current and hence, well-behaved transistor performance. In addition, SOI structure suppresses the junction leakage problem, resulting in good agreement between the source current and drain current of the MOSFET.

INTRODUCTION

With continuous scaling of the Si based MOSFET (Metal Oxide Semiconductor Field Effect Transistor) reaching its fundamental limits, alternative substrates have been considered to replace Si [1]. Among them, thin SGOI with high percentage Ge concentration has many attractive properties: 1) SiGe with high percentage Ge provides higher intrinsic electron and hole mobility than that of Si [2]; 2) thin body structured transistor can minimize the off leakage current, reduce the parasitic junction capacitance and provide better scalability to shorter channel length [1]. However, two main challenges exist for high Ge percentage thin SGOI MOSFET. First, high Ge concentration thin SGOI substrate fabrication is a challenging task. Techniques for SGOI layers fabrication such as the SIMOX (Separation by implanted oxygen technique) [3], and the wafer bonding techniques [4], have been demonstrated. However, SIMOX is not suitable for SGOI fabrication with Ge concentration higher than 30% [3], and wafer bonding technique is difficult to achieve thin SGOI. In addition, the significant increase of series resistance without S/D (source/drain) extension is a serious problem for thin body transistor, which results in the reduction of drive current [5].

Recently, condensation method to fabricate SGOI has drawn great attention and excellent results have been demonstrated [6]. Rather than oxidizing single crystal SiGe on SOI prepared by epitaxial growth, we have proposed cost-effective and simple SGOI fabrication technique by oxidizing amorphous SiGe film on SOI substrates [7], which is also promising for high Ge percentage thin SGOI substrate fabrication. On the other hand, in order to reduce the S/D resistance, Schottky S/D technology has been proposed as an alternative to heavily doped S/D, and promising results have recently been reported, including Schottky S/D transistors on Si-bulk and Ge-bulk substrates using various metals (Ni, Pt, Er, Yb, Dy) [8-10]. Combination of ultra-thin body structure and Schottky S/D engineering has been considered as one of the promising non-classical CMOS technology. Recently, Pt-germanide Schottky S/D transistor on Ge-on-insulator using buried SiO_2 as gate dielectric and Si-substrate as bottom gate electrode has been reported with excellent performances, operating in accumulation mode [11]. In this report, we successfully demonstrate the Schottky S/D transistor using Ni-germanosilicide on the thin SGOI wafer and HfO_2/TaN gate stack fabricated by conventional self-aligned top gate structure.

EXPERIMENTAL DETAIL

For preparation of SGOI, we achieved SGOI (SiGe thickness = $\sim$30 nm) with Ge atomic concentration of 65% by oxidization of amorphous SiGe film on single crystal SOI wafer. Low Ge content amorphous SiGe film was deposited on single crystal SOI wafer by co-sputtering of pure Si and Ge targets. The oxidation process is first done at a higher temperature of 1050°C until the Ge percentage in the film reaches $\sim$50%, and after removing the top oxide layer by DHF clean, further oxidation is carried out at a lower temperature of 900°C to ensure that the film will not melt during the whole oxidation process. Detail process flow can be found in reference [7]. Single crystalline of a similar film prepared by the same method was investigated and verified by the by Raman, XRD (X-Ray diffraction) and fast fourier transform [7].

After dipping in dilute HF to remove the native oxide, the $Si_{0.35}Ge_{0.65}OI$ substrate was loaded into a sputter chamber with a base pressure of 5E-7 Torr. A thin ($\sim$5 Å) amorphous Si passivation layer was deposited on SGOI to suppress the GeO_2 formation during the subsequent high-k formation step, which causes the degradation of MOS devices [13], followed by pure Hf metal deposition by DC sputtering. Rapid thermal oxidation of the sputtered Hf-layer on SGOI was done at 500°C in a N_2/O_2 ambient. A 150 nm TaN metal gate was deposited by reactive sputtering of Ta target in N_2 ambient. After the gate pattering, Si_4N_3 spacer was deposited by plasma enhanced chemical vapor deposition. Immediately after spacer etching and DHF dipping to remove the oxide in the S/D area, pure Ni was deposited on the SGOI wafer by sputtering. The Ni-germanosilicide on the S/D was formed at 400°C for 1 minute in N_2 ambient at an atmosphere pressure. Finally, unreacted Ni residue was removed by HNO_3 solution.

DISCUSSION

Fig.1 (A) illustrates the single crystal SGOI substrate fabrication process. Amorphous SiGe was deposited on single crystal SOI wafer by co-sputtering Si and Ge targets, after multi-step oxidation process, single crystal SGOI wafer was successfully fabricated. Fig. 1 (B) shows the high resolution TEM (Transmission Electron Microscopy) of the achieved SiGe layer on insulator used for Schottky S/D transistor fabrication. The thickness of the single crystal SiGe layer is found to be $\sim$30 nm with Ge concentration of $\sim$65% detected by EDX (Electron

Dispersive X-ray). Ge concentration across the film is quite uniform within this ~30 nm SiGe layer.

High resolution TEM of the gate stack is shown in Fig. 2. The amorphous interfacial layer is quite uniform and that layer is believed to be the compound of Si [14]. As can be seen in high resolution TEM of gate stack in Fig. 2, a conformal HfO_2 film on single crystal $Si_{0.35}Ge_{0.65}$ channel remained amorphous after transistor fabrication owing to the low temperature Schottky S/D formation process. Ultra thin interfacial layer is observed between HfO_2 film and $Si_{0.35}Ge_{0.65}$ channel, which is believed to be formed during oxidation process. In addition, the clear smooth interface between $Si_{0.35}Ge_{0.65}$ channel and Ni-germanosilicide is observed at S/D region. The FFT image of the SiGe is shown in the inset of the Fig. 2 which confirms its single crystal nature. The epitaxial growth of Ni-germanosilicide grain on $Si_{0.35}Ge_{0.65}$ (001) is also observed which is verified by the FFT image from the Ni-germanosilicide region shown in inset.

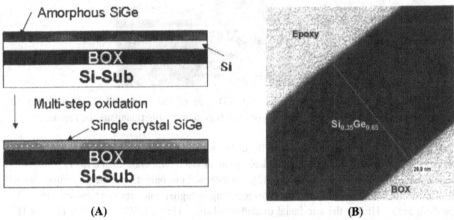

(A)   (B)

Fig. 1 (A) Single crystal SGOI fabrication process; amorphous SiGe was deposited on SOI wafer by co-sputtering Si and Ge targets; (B) High resolution TEM image of the SGOI wafer prepared by oxidation of amorphous SiGe on SOI wafer. The thickness of the body is ~30nm and the Ge atomic concentration detected by EDX is 65%.

The sufficient overlap of Ni-germanosilicide S/D with gate electrode may be explained by the fact that Ni is the main diffusing species during solid-state reaction of Ni with Si and Ge [15]. The atomic composition of Ni: Si: Ge detected by EDX is 52:22:26. According to the previous report, three stable ternary phases, Ni_3Si-Ni_3Ge, Ni_2Si-Ni_2Ge, and NiSi-NiGe can be formed during reaction of Ni with SiGe [15]. And at a 400℃ rapid thermal anneal process, our EDX data indicates that NiSi-NiGe is the preferred phase [15].

Fig. 3 shows the inversion C-V (capacitance-voltage) curve of the fabricated SGOI p-MOSFET measure at 1 MHz, the peak inversion capacitance corresponds to an equivalent oxide thickness of 2.2 nm considering the quantum mechanical effect. The inset shows gate leakage current density as a function of the gate bias when the channel is in inversion. A gate leakage of $2x\ 10^{-4}$ A/cm^2 at a

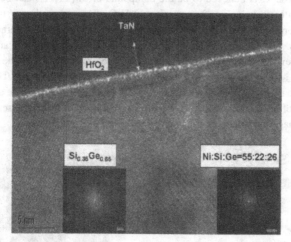

Fig. 2: TaN-HfO$_2$-Si$_{0.35}$Ge$_{0.65}$ gate stack and S/D area of the fabricated SGOI Schottky S/D transistor; insets ate FFT images from Si$_{0.35}$Ge$_{0.65}$ region and Ni-germanosilicide region.

gate bias of -1 V is achieved. MOSFETs made on Ge or SiGe (with high percentage of Ge) substrates usually suffer from high gate leakage at elevated temperature, due to: 1) formation of poor quality Ge oxide interfacial layer [13], which will not only distort the C-V curve but will also lower the band offset to the substrates resulting in higher gate tunneling current [16, 17], 2) the diffusion of Hf into the interfacial oxide/Ge-substrate [18], 3) diffusion of Ge into the HfO$_2$ film [19]. Hence, high temperature S/D activation process should be avoided during the fabrication of Ge and SiGe (with high Ge concentration) MOSFET. By using low temperature process for Ni-germanosilicide S/D formation, we successfully demonstrate Schottky S/D p-MOSFET on SGOI with excellent C-V characteristic together with low gate leakage.

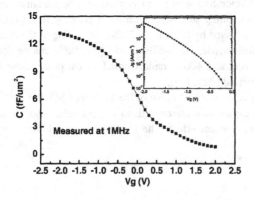

Fig. 3: Inversion capacitance-voltage curve of the fabricated SGOI p-MOSFET. Gate leakage density is shown in inset.

Fig.4 shows the typical drain current (I_d) and source current (I_s) versus drain voltage (V_d) curves of thin SGOI Schottky S/D PMOSFET with channel width/length= 400/5 um. Smaller band gap of Ge and SiGe (with high Ge concentration) is the cause of high junction leakage of MOSFET made on these substrates [20]. In addition, the difficulties in achieving highly doped S/D due to the poor activation and solid solubility limits of dopants in Ge-based MOSFET cause junction leakage and difference in Is and Id [16, 21, 22]. Moreover, the ambipolar carrier conduction at the S/D edges for Schottky S/D transistor induces higher current at drain [4]. Ni is chosen as the Schottky metal electrode in that the reaction of Ni with SiGe leads Ni-germanosilicide, which is expected to provide low Schottky barrier for hole according to the measured Schottky barrier height of NiGe on n-type Ge substrate [23]. On the other hand, the large barrier for electrons will suppress the electrons tunneling from drain into channel under high drain bias voltage resulting in reduced junction leakage. Observed excellent agreement between I_d and I_s of fabricated SGOI Schottky S/D transistor indicates good interface at Schottky contacts, and low junction leakage. Extracted peak hole mobility is ~ 120 cm^2/V-s at low electric field, which is comparable with hole mobility from conventional bulk Ge p-MOSFET with HfO$_2$/TaN gate stack [16].

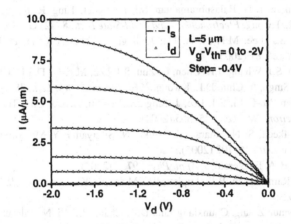

Fig. 4: Typical I_d/I_s-Vg characteristics for the fabricated SGOI Schottky S/D p-MOSFET transistor.

CONCLUSION

In summary, we fabricate and demonstrate the Schottky S/D transistor using Ni-germanosilicide on thin SGOI wafer prepared by novel and cost effective approach of oxidizing co-sputtered amorphous SiGe film deposited on SOI substrates, and HfO_2/TaN gate stack integrated with conventional self-aligned top gate struc ure. Low Ni-germanosilicide S/D formation temperature leads to the excellent inversion CV characteristics and low gate leakage density. Excellent transistor output characteristics are presented. Extracted peak hole mobility is comparable with hole mobility from conventional bulk Ge p-MOSFET with HfO_2/TaN gate stack.

Reference:

1. Y.-K Choi, Y.-C Jeon, P. Ranade, H. Takenuchi, T.-J King, J. Bokor, C.M. Hu, *58th Device Research Conference 2000*, **23**, (2000).

2. M.L. Lee and E.A. Fitzgerald, *J. Appl. Phys. Lett.*, **97**, 101 (2005).

3. S. Fukatsu, Y. Ishikawa, T. Saito, and N. Shibata, *Appl. Phys. Lett.* **72**, 3485 (1998).

4. G. Taraschi, A.J. Pitera, and E.A. Fitzgerald, *Solid-State Elecıron.* **48**, 1297 (2004).

5. International Technology Roadmap for Semiconductors, 2004 update, *Process Integration, Devices, and Structures*, (2004).

6. T. Tezuka, N. Sugiyama, and S. Takagi, *Appl. Phys. Lett.* **79**, 1789 (2001).

7. F. Gao, S. Balakumar, N. Balasubramanian, S.J. Lee, C.H. Tung, R. Kumar, T. Sudhiranjan, Y.L. Foo, and D.-L. Kwong, *Electronchem. and Solid-State Lett.*, **8** (12) G337-G340 (2005).

8. S. Zhu, R. Li, S.J. Lee. M.F. Li. A. Du. J. Singh, C. Zhı, A. Chin, D. L. Kwong, *IEEE Electron. Dev. Lett.*, **26**, 81 (2005).

9. S. Zhu, H.Y. Yu, S.J. Whang, J.H. Chen, C. Zhu, S.J. Lee, M.F. Li. D.S.H. Chan, W.J. Yoo. A. Du, C.H. Tung, J. Singh, A. Chin, D.L. Kwong, *IEEE Electron. Dev. Lett.*, **25**, 268 (2004).

10. S. Zhu, J. Chen, M.-F. Li, S.J. Lee, J. Singh, C.X. Zhu, A. Du, C.H. Tung, A. Chin, D.L. Kwong, *IEEE Electron. Dev. Lett.*, **25**, 565 (2004).

11. T. Maeda, K. Ikeda, S. Nakaharai, T. Tezuka, N. Sugiyama, Y. Moriyama, and S. Takagi, *IEEE Electron. Dev. Lett.*, **26**, 102 (2005).

12. M. L. Lee and E. A. Fitzgerald, *J. Appl. Phys.*, **97**, 1 (2005).

13. C.O. Chui, S. Ramanathan, B.B. Triplett, P.C. Mclntyre, K.C. Saraswat, *IEEE Electron. Dev. Lett.*, **23**, 473 (2002).

14. Nan Wu, Qingchun Zhang, Chunxiang Zhu, DSH Chan, M.F Li, N. Balasubramanian, Albert Chin, and Dim-Lee Kwong, *Appl. Phys. Lett.*, **85**, 4127 (2004).

15. S.L. Zhang, M. Ostling, **Critical Reviews in Solid State and Mater. Sci.**, *28*, 1, (2003).

16. S.J. Whang, S.J. Lee, F. Gao, N. Wu, C.X. Zhu, J.S. Pan, L J. Tan, D.L. Kwong, *Tech. Dig. –Int. Electron Devices Meet.* 2004, **307** (2004).

17. V.V Afanas'ev, A. Stesmans, *Appl. Phys. Lett.* **84**. 2319 (2004).

18. H. Kim, P.C. Mclntyre, C.O. Chui, K.C. Saraswat, M.H. Cho, *Appl. Phys. Lett.*, **85**, 2902 (2004).

19. S.V. Elshocht, B. Brijs, M. Caymax, T. Conard, T. Chiarella, S.D. Gendt, B.D. Jaeger, S. Kubicek, M. Meuris, B. Onsia, O. Richard, I. Teerlinck, J.V. Stıenbergen, C. Zhao, M. Heyns, *Appl. Phys. Lett.*, **85**, 3824 (2004).

20. C. K. Maiti, N. B. Chakrabarti and S. K. Ray, **Strained Silicon and Heterostructures:**

materials and devices, (The insititute of Electrical Engineers, 2001).

21. Y.S. Suh, M.S. Carroll, R.A. Levy, M.A. Sahiner, G. Bisognin, C.A. King, *IEEE Trans. Electron Devices*, **52**, 91 (2005).

22. A. Nayfeh, C.O. Chui, T. Yonehara, K.C. Saraswat, *IEEE Electron. Dev. Lett.*, **26**, 311 (2005).

23. D.Z Chi, R.T. P. Lee, S.J. Chua, S.J. Lee, S. Ashok and D.-L. Kwong, *Appl. Phys. Lett.* **97** 113706 (2005).

Mater. Res. Soc. Symp. Proc. Vol. 913 © 2006 Materials Research Society   0913-D01-07

Schottky-Barrier Height Tuning Using Dopant Segregation in Schottky-Barrier MOSFETs on Fully-Depleted SOI

Joachim Knoch, Min Zhang, Qing-Tai Zhao, and Siegfried Mantl
Institute of Thin Films and Interfaces, ISG1, Research Center Juelich, Juelich, D-52425, Germany

ABSTRACT

In this paper we demonstrate the use of dopant segregation during silicidation for reducing the effective potential barrier height in Schottky-barrier metal-oxide-semiconductor field-effect-transistors (SB-MOSFETs). N-type as well as p-type devices are fabricated with arsenic/boron implanted into the device's source and drain regions prior to silicidation. During full nickel silicidation a highly doped interface layer is created due to dopants segregating at the silicide-silicon interface. This doped layer leads to an increased tunneling probability through the Schottky barrier and hence leads to significantly improved device characteristics. In addition, we show with simulations that employing ultrathin body (UTB) silicon-on-insulator and ultrathin gate oxides allows to further improve the device characteristics.

INTRODUCTION

Schottky-barrier metal-oxide-semiconductor field-effect transistors (SB-MOSFETs) have recently attracted a renewed interest as an alternative over conventional MOSFETs since they offer solutions for aggressively scaled devices related to the source and drain contacts [1,2]: Due to metallic contacts directly attached to the channel, SB-MOSFETs exhibit low extrinsic parasitic resistances, offer easy processing and allow for well-defined device geometries down to smallest dimensions. However, at a metal-semiconductor contact a SB always builds up which is usually much larger than a few $k_B T$ and hence limits the on-state performance and causes poor subthreshold behavior. Recently, dopant segregation (DS) during silicidation has been successfully used in n-type bulk SB-MOSFETs and a significant reduction of the effective Schottky-barrier height and hence a strong improvement of the electrical characteristics of SB-MOSFETs could be achieved [3]. Here we will show that arsenic as well as boron segregation during silicidation can be used to strongly improve the electrical characteristics of both, n-type as well as p-type SB-MOSFETs on fully depleted SOI with fully nickel silicided source and drain contacts. A further increase of the on-state current can be achieved by employing ultrathin body (UTB) SOI and ultrathin gate oxides which leads to a better carrier injection through the SBs.

EXPERIMENT

Commercially available SOI wafers with a p-type doping of 10^{15} cm^{-3} were first thinned by a cycle of dry/wet thermal oxidation and diluted HF stripping to a thickness of 25nm. After a standard mesa isolation and RCA cleaning a ~3.5nm thick gate oxide is grown followed by the deposition of 200nm n-doped poly-silicon. Subsequently the poly-silicon gate is patterned using reactive ion etching. At this point, source and drain areas of one part of the samples were implanted (Fig. 1 (a)) with arsenic and boron with an energy of 5keV and a dose of $5 \cdot 10^{14} \text{cm}^{-2}$ (arsenic) and an energy of 2keV and a dose of $3 \cdot 10^{15} \text{cm}^{-2}$ (boron), respectively. Control samples were left without any implantation as a reference. Next, gate spacers were generated with LPCVD grown SiO_2 and nickel was deposited by e-beam evaporation. For the subsequent silicidation step, a temperature of 450° C and a time of 20s (30s) was chosen for the arsenic (boron) implanted devices in order to facilitate the encroachment of NiSi into the channel region as can be seen in the cross-sectional TEM image in Fig. 1. This is a simple way to ensure that the silicide-silicon interface is well in the gate region without the need for ultrathin spacers. Furthermore, it is ensured that dopants segregate at the contact-channel interface. At the same time, the silicidation temperature is low enough to avoid any thermally activated dopant diffusion during the silicidation step. Secondary ion mass spectroscopy investigations of bulk test samples (not shown here) show that a dopant concentration $> 10^{20} \text{cm}^{-3}$ ($> 4 \cdot 10^{19} \text{cm}^{-3}$) is found at the interface in case of arsenic (boron) [4].

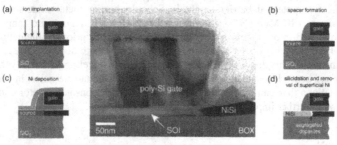

Fig. 1: Schematics of the device fabrication process (a) – (d). The TEM image shows a cross-section of a readily fabricated device, showing the NiSi encroachment in the gate area.

EXPERIMENTAL RESULTS

Figure 2(a) shows transfer characteristics of a SB-MOSFET with arsenic segregation. Asymmetric ambipolar operation can be observed with a significantly larger current level n-type branch and an inverse subthreshold slope of the n-type branch of $S_{ii)} = 70\,\text{mV/dec}$, close to the thermal limit. Figure 2(b) shows transfer characteristics of a SB-MOSFET device with boron segregation. A similar asymmetric ambipolar behavior can be observed as in case of the arsenic device, the only difference being that the device acts as a p-type transistor. The inverse subthreshold slope of the p-type branch is 65mV/dec, again close to the thermal limit. The inverse subthreshold slopes of both devices suggest that the off-state of the transistors is determined by thermal emission as in a conventional MOSFET rather than tunneling despite the high SB of 0.64eV (0.46eV) for electrons (holes) at the NiSi-silicon interface. As will become clear below the high dopant concentration in the segregation layers results in a strongly reduced effective Schottky barrier height leading to an off-state similar to a conventional device. Looking at the output characteristics as shown in the insets of Fig. 2(a) and (b) a difference between the arsenic and the boron device becomes apparent. While the arsenic device shows a linear increase of current for small bias typical of a conventional MOSFET the boron device exhibits the exponential increase of current for small drain-source voltages usually observed in SB-MOSFETs. The reason for this is that in case of the boron device the dopant concentration in the segregation layer is significantly smaller as compared to the arsenic device [4]. As a consequence, the effective SB is decreased less effectively and hence the off-state of the boron device is determined by thermal emission as indicated by a slope of 65mV/dec whereas the on-state still shows the typical SB-MOSFET behavior.

The effect of, e.g., arsenic segregation on the electrical characteristics of SB devices can be understood by looking at the conduction band profile at the source channel interface for various gate voltages as depicted in Fig. 3(a). Due to the formation of a highly doped segregation layer a strong band bending occurs, depending on the type and concentration of doping. In case of the arsenic device a high As-concentration allows for a large tunneling probability of electrons through the SB. In other words the strong bending makes the SB very thin such that while the true SB height remains unaltered the effective Schottky barrier for thermal emission appears to be lowered. Consequently, in the device's off-state (corresponding to the gate voltage range denoted '1' and '2' in Fig. 3(a)) the charge carrier flow is determined by the bulk potential in the channel (as in a conventional MOSFET) in contrast to the tunneling dominated carrier injection in case of an SB-MOSFET. In the on-state, however, the current is determined by the tunneling through the lowered effective SB ('3' and '4'). As the SB is decreased for electrons it is increased for holes yielding a suppression of the hole leakage current and hence to asymmetric ambipolar characteristics. In principle, the same is true in case of boron doping: Holes can easily be injected into the channel through the thinned SB at the contact channel interface. However, the exponential increase of the output characteristics for small bias indicates that a substantial barrier is still present at the silicide/silicon interface as has already been mentioned above.

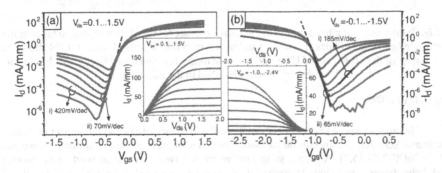

Fig. 2: Transfer characteristics of devices with arsenic (a) and boron (b) segregation. The insets show the corresponding output characteristics.

Looking at the linear output characteristics and the almost ideal subthreshold behavior of the arsenic device, for instance, one could argue that the SB-MOSFET with arsenic segregation is a conventional MOSFET. It is important, however, to note that there is a difference between a conventional MOSFET and the present SB-MOSFETs with dopant segregation. This difference can be inferred from the leakage current, i.e. from the p-type branch of the n-type device (Fig. 2(a)) and the n-type branch of the p-type device (Fig. 2(b)), respectively. Although the ambipolar characteristics are asymmetric the leakage is significantly higher than in a conventional MOSFET. This shows that the dopant segregation layer, i.e. the layer where the carrier concentration is approximately $10^{20}\,\mathrm{cm}^{-3}$ has to be on the order of a few nanometers only, consistent with [3] and with simulations presented in [6]. The reason is that for increasing segregation length the potential barrier for holes/electrons (in the n-type/p-type device) becomes broader and hence the leakage current gets more and more suppressed. For large segregation lengths the transistor would eventually behave like a conventional device since in this case the segregation layer acts as the actual, doped source/drain contact. Therefore, the SB-MOSFETs with dopant segregation are SB-MOSFETs with a substantially reduced Schottky barrier rather than a conventional MOSFET. As a result, using a silicide with a low SB and combining this with DS allows to fabricate devices with an intrinsic performance close to a conventional device still offering low extrinsic parasitic resistances and a good scalability.

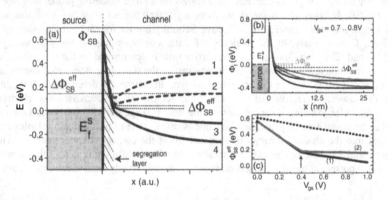

Fig. 3: (a) Conduction band profile at the source contact for two different gate voltages in the off-state (gray dashed lines '1' and '2') and the on-state (black lines '3' and '4'). (b) shows the thinning of the SB in case of UTB SOI and gate oxide. (c) shows the effective SB height for different devices described in the main text.

SIMULATIONS

In order to gain a better understanding of the influence of DS and the impact of differing oxide and SOI body thicknesses on the device behavior we have performed quantum simulations of SB-MOSFETs [5,7]. A proper simulation requires a two-dimensional solution of Poisson's and Schrödinger's equation. However, it is possible to describe the electrostatics of fully depleted SOI devices by a modified, 1D Poisson equation [8] given by

$$\frac{d^2\Phi_f}{dx^2} - \frac{\Phi_f - \Phi_g + \Phi_{bi}}{\lambda^2} = \frac{e\left(\rho(x) + N_{seg}\right)}{\varepsilon_0 \varepsilon_{si}} \tag{1.1}$$

where Φ_f, Φ_g and Φ_{bi} are the surface potential, the gate potential and the built-in potential, respectively; $\lambda = \sqrt{\varepsilon_{si}/\varepsilon_{ox} d_{si} d_{ox}}$ is the relevant length scale on which potential variations are being screened. The effect of DS is accounted for by a step function like doping profile of spatial extension l_{seg} and concentration N_{seg} at the contact-channel interfaces. The charge $\rho(x)$ in and current through the channel is calculated using the non-equilibrium Green's function formalism [9] where the equation for the charge is solved self-consistently with eqn. (1.1). Ballistic transport is assumed in order to give an upper estimate of the possible device performance. More details on the calculations can be found elsewhere [5,7].

To verify the interpretation of a performance improvement due to an effective SB lowering and to investigate the impact of a varying body and gate oxide thickness on the transistor performance we have simulated transfer characteristics of SB-MOSFETs with DS ($N_{seg} = 2 \cdot 10^{20} \text{cm}^{-3}$, $l_{seg} = 2 \text{nm}$) and a fixed SB of 0.64eV for two different body and oxide thicknesses, namely (1) d_{ox}=1nm, d_{si}=5nm and (2) d_{ox}=5nm, d_{si}=25nm. A channel length of L=60nm in case of (1) and L=160nm in case of (2) was found to be sufficient to ensure long-channel behavior. An effective SB height Φ_{SB}^{eff} can be determined from the self-consistent calculations by associating the barrier with an energy where the transmission through the SB has dropped to a specific value. Figure 3(c) shows Φ_{SB}^{eff} as a function of gate voltage. The one-to-one change of Φ_{SB}^{eff} with V_{gs} for small gate voltages (off-state) reflects the fact that Φ_{SB}^{eff} is smaller than the bulk potential in the channel meaning that in this V_{gs}-range the device behaves like a "bulk switching" transistor in both device cases, (1) and (2). This can also be inferred from the conduction band profiles denoted with '1' and '2' in Fig. 3(a) showing that in the off-state the bulk potential plays the role of Φ_{SB}^{eff} which leads to an almost ideal off-state. However, for large gate voltages, i.e. in the device's on-state, Φ_{SB}^{eff} changes much slower with V_{gs} as indicated in Fig. 3(a). In case of device (2) it remains nearly constant whereas in case (1), Φ_{SB}^{eff} continuously decreases with increasing V_{gs} due to a much better gate control of the potential distribution of the SB in case of ultrathin bodies and oxides. This improved gate control can be inferred from the conduction band profile at the source contact for two different V_{gs} as shown in Fig. 3(b): although the potential is equal for the different devices far away from the contact-channel interface, the improved gate control in case of UTB SOI and ultrathin gate oxides leads to a much thinner SB in the device's on-state. This means that in case of (1) a much better on-state can be expected. For comparison, we also plot Φ_{SB}^{eff} for a device of type (1) but without DS (black dotted line in Fig. 3(c)). The curve exhibits the same slope as the device with DS but begins at a much larger SB as indicated by the arrows. Thus, DS strongly reduces the effective SB height.

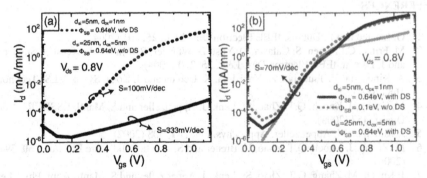

Fig. 4: (a) Transfer characteristics of SB-MOSFETs without DS. (b) shows a comparison of devices with DS with a device without DS but much lower SB height. Parameters are as indicated in the figure.

Figure 4(a) shows transfer characteristics of two devices of type (1) and (2) without DS. A clear improvement in case of the device of type (1) can be seen which is due to the better gate control leading to an increased carrier injection [5]. Figure 4(b) shows the transfer characteristics of two devices of type (1) (black line) and (2) (gray line) with DS along with a device of type (1) (gray dotted line) without DS but with a much smaller SB of 0.1eV. Both devices with DS exhibit an almost ideal off-state showing that DS has effectively lowered the Schottky barrier height. However, the on-state is significantly different with a much larger on-current in case of (1) with UTB and ultrathin gate oxide as anticipated from the discussion above and the result shown in Fig. 4(a). Owing to the increased transmission through the SB, the device of type (1) with DS and $\Phi_{SB} = 0.64\,\mathrm{eV}$ exhibits approximately the same on- and off-state performance as the device without DS but with a barrier of 0.1eV only. As a result, the DS technique greatly relaxes the requirements for low Schottky barrier electrode materials due to an increased tunneling probability through the SB. In combination with UTB SOI and ultrathin gate oxides an even further improved on-state can be achieved allowing for high performance SB-MOSFET devices.

CONCLUSIONS

In conclusion, we studied the impact of dopant segregation during silicidation on the performance of SOI SB-MOSFETs. It was shown that a highly doped interface layer results in a drastic reduction of the effective Schottky barrier height for both, n- as well as p-type devices. Experimental devices exhibited a significantly improved on- and off-state with an inverse subthreshold slope close to the thermal limit. Simulations show that the use of UTB SOI and ultrathin gate oxides enables a further improvement of the transistor's on-state due to a more substantial lowering of the effective Schottky barrier height. Consequently, DS and ultrathin body/oxide devices allow to combine an excellent intrinsic performance with the specific advantages of SB-MOSFETs.

REFERENCES

1. G. Larrieu and E. Dubois, IEEE Electron Dev. Lett., **25**, 801 (2004).
2. M. Fritze, C.L. Chen, S. Calawa, D. Yost, B. Wheeler, P.Wyatt, C.L. Keast, J. Snyder and J. Larson, IEEE Electron Dev. Lett., **25**, 220 (2004).
3. A. Kinoshita, Y. Tsuchiya, A. Yagishita, K. Uchida and J. Koga, Symp. VLSI Technol. 168 (2004).
4. M. Zhang, J. Knoch, Q.T. Zhao, St. Lenk, J. Appenzeller and S. Mantl, ESSDERC Conf. Digest, 457 (2005).
5. J. Knoch and J. Appenzeller, Appl. Phys. Lett., **81**, 308 (2002).
6. M. Zhang, J. Knoch, Q.T. Zhao, U. Breuer and S. Mantl, Solid-State Electron., **50**, 594 (2006).
7. J. Knoch, M. Zhang, Q.T. Zhao, St. Lenk, J. Appenzeller and S. Mantl, Appl. Phys. Lett., **87**, 263505 (2005).
8. K. Young, IEEE Trans. Electron Dev., **36**, 399 (1992).
9. S. Datta, *Electronic Transport in Mesoscopic Systems*, Cambridge Univ. Press (1998).

Mater. Res. Soc. Symp. Proc. Vol. 913 © 2006 Materials Research Society   0913-D01-09

Visualisation of Ge Condensation in SOI

Kristel Fobelets[1], Benjamin Vincent[1], Munir Ahmad[1], Astolfi Christofi[2], and David McPhail[2]

[1]Electrical and Electronic Engineering, Imperial College London, Exhibition Road, London, London, SW7 2BT, United Kingdom

[2]Materials, Imperial College London, Exhibition Road, London, London, SW7 2BT, United Kingdom

ABSTRACT

We use a novel technique CABOOM – Characterisation of Alloy concentration via Beveling, Oxidation and Optical Microscopy – to visualize the change of the Ge concentration sandwiched between two SiO_2 layers during the Ge condensation process. CABOOM is very sensitive to variations in the gradient of the Ge concentration in the SiGe layer and thus gives a fast and simple way to interpret the condensation process. We present a systematic study of Ge condensation in a 120nm thick $Si_{0.92}Ge_{0.08}$ layer on a 60 nm Si body SOI (silicon-on-insulator) as a function of oxidation temperature and time, using CABOOM, SIMS and XRD. CABOOM shows the non-linear variation of the Ge diffusion as a function of process time.

INTRODUCTION

In the quest for fast and low power integrated circuits, silicon-on-insulator (SOI) substrates have shown speed and power performance improvements. The introduction of Si:SiGe heterojunctions have also led to device and circuit improvements beyond those possible in bulk Si. Research is now aimed at combining both technologies into SiGe-on-insulator (SGOI), strained-Si-on-insulator (sSOI) and Ge-on-insulator (GOI).

The two main fabrication techniques are SIMOX [1] and SMARTCUT [2]. Both techniques are based on epitaxial growth of virtual substrates and relaxed SiGe buffer layers, making production time-consuming and expensive. They suffer both from limitations on maximum Ge concentration ($\pm 30\%$) due to dislocation formation and temperature restrictions. In [3] another fabrication method was proposed: direct epitaxial growth of low Ge contents SiGe on SOI followed by Ge condensation between two oxide layers. This led to higher Ge concentrations, beyond the as-grown value and even to pure GOI substrates [4].

Ge condensation is based on two competitive mechanisms during dry thermal oxidation of SiGe grown on SOI. On one hand the SiGe alloy is oxidized which causes a Ge accumulation at the interface between the alloy and the top oxide that is being formed. No GeO_2 is produced at the interface because of the preferential formation of SiO_2 at high temperatures [5]. On the other hand Ge can diffuse into the remaining SiGe/Si layers while diffusion into the substrate is blocked by the buried oxide (BOX) of the SOI wafer. These mechanisms depend on the oxidation temperature [6]. For temperatures below 1200°C the diffusion length of Ge in Si is smaller than the diffusion length of O_2 in SiO_2, thus oxidation is the dominant mechanism and Ge diffusion is minimal, this tends to accumulate Ge at the SiGe/SiO_2 interface. For temperatures above 1200°C the Ge diffusion length in Si is larger than O_2 in SiO_2, thus Ge diffusion is enhanced and more homogeneous Ge profiles are expected or loss of Ge occurs before the formation of a sufficiently thick top SiO_2 layer. One has to keep in mind that increasing the Ge concentration lowers the melting point of SiGe, limiting the maximum temperature [7]. The quality of the SGOI structure (homogeneity of the alloy, strain inside the alloy, dislocation formations...) obtained by this procedure depends on the oxidation time and temperature.

We have carried out dry thermal oxidation within a wide temperature range (1050-1200°C) and different oxidation times (2-8hrs) on a 60nm body SOI layer covered with 120nm $Si_{0.92}Ge_{0.08}$. All oxidations were carried out in a pure O_2 environment and with the same ramp times. CABOOM [8] has been used to give information on the homogeneity of the Ge concentration in the SGOI layer. SIMS and XRD measurements are added to support the CABOOM results.

CABOOM technique

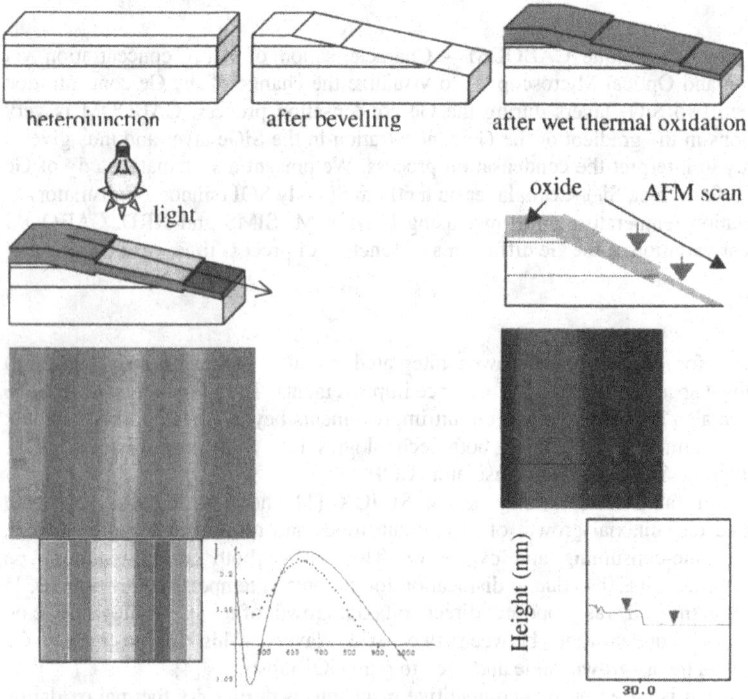

Figure 1: Top 3 figures show beveling and wet oxidation result (grey=SiO_2). Left figures give the view under an optical microscope and the reconstruction of the colors (reflection spectrum R-λ). Right figures give the steps in oxide thickness measured via AFM.

CABOOM visualizes the Ge profiles in SiGe heterostructures in a very simple way [8]. The process relies on the fact that the *steam* oxidation rate of SiGe is dependent on the Ge concentration in contrast to dry oxidation. As shown in fig.1 beveling will present all the SiGe layers in the heterojunction at the surface. When this surface is oxidized in a low thermal steam

process then the surface presents steps in oxide thickness directly related to the Ge concentration in the layer underneath. Reflection of light is very sensitive to the SiO_2 thickness and thus interference presents the different layers as different colors. We have optimized our CABOOM simulator presented in [8] to reconstruct the colors from multi-layers visualized by

Figure 2: Left top: microscope picture of the as-grown wafer (before wet thermal oxidation) visualizes the BOX and top layer thickness. Right top: after wet thermal oxidation, visualizes the oxide thickness gradient and thus contains the Ge concentration information. Bottom figures are the reconstructed colors of the CABOOM simulator.

our in-house optical microscope. The simulator is calibrated using a beveled SiO_2 layer of known thickness. The limitations of CABOOM are the need for smooth beveled samples and a minimum layer thickness of 30 nm. For homogeneous layers with Ge concentrations $x_{Ge}<56\%$, a x_{Ge} difference of 1.2% can be reconstructed visually. The use of CABOOM in the situation where the SiGe layers are on top of BOX is more complicated as the BOX already imposes interference colors (see fig.2 left). Although this makes estimation of the effective Ge concentration difficult (lowers the resolution to 5%), it is still an easy tool to visualize the condensation process (see fig.2 right).

CABOOM applied to Ge condensation in SOI

Fig.3 shows the CABOOM results with a bevel angle of 20' and wet oxidation at 700°C for 2 hrs on 3 different samples: as-grown, dry oxidized at 1100°C for 2 and 5 hrs.

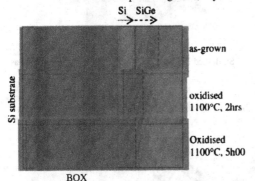

Figure 3: CABOOM visualizes the Ge condensation on SOI. The red box indicates the BOX. The dotted line gives the Si/SiGe interface and the dashed line the SiGe/SiO₂ interface

As the BOX does not change during dry oxidation nor the CABOOM preparation, the color variation through its thickness remains constant for all samples and thus can be used to initialize the start of the thin Si/SiGe film in all samples. The colors on the beveled surface of the steam oxidized samples are directly related to the thickness of the top oxide and not to the Si/SiGe layer thickness underneath as a consequence of the small refractive index of SiO_2 compared to Si or SiGe. This means that color changes between the right edge of the BOX and the dashed line in fig.2 are due to inhomogeneities in the Ge concentration in the top Si/SiGe layer.

The microscope picture in the as-grown sample shows the Si layer as a constant color while the CVD grown SiGe layer already shows a non-uniform Ge profile mainly due to Ge segregation during growth. The final color is consistent with a Ge concentration of 13%±2% as determined by CABOOM spectral analysis (see fig.1 left bottom). After dry oxidation at 1100°C for 2hrs, it is clear that an almost pure thin Si layer remains, while a steep Ge concentration increase is noticed between the dotted and dashed line. An estimation of the Ge concentration at the top edge of the SiGe gives approximately 35%±2%. The remaining Si layer disappears completely when the oxidation time is increased to 5 hrs, giving an almost homogeneous SiGe with an estimated Ge concentration of ±27%±5%.

Condensation was performed again at 1050°C for different times. CABOOM was done at 700°C for only 1 hour to avoid possible Ge diffusion in this step. The results are presented in fig.4 together with SIMS. After 2 hours of dry oxidation, consumption of the SiGe layer happened but negligible Ge diffusion, due to the low T condensation process. A high Ge peak is seen at the Si/SiGe interface, consistent with SIMS (see red arrows). This pile up is absent at higher temperatures (fig.3). Only after 3 hrs of condensation does the Ge start to diffuse quickly in the narrowing layer. After 4hrs, the Si layer is almost entirely removed. For longer dry oxidation times, t=8hrs, the top oxide formed during the dry oxidation process and the BOX layer of the SOI are very close, making the derivation of the bevel edge almost impossible. A shallower bevel or an adapted CABOOM technique that protects the top surface from oxidation would allow better interpretation for thin layers. For further increased oxidation times SIMS fails to detect any remaining Ge indicating that the condensation process is not self-limiting.

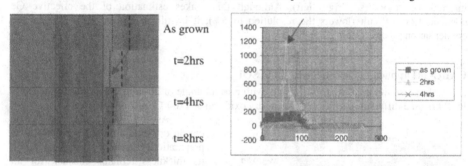

Figure 4: Left: CABOOM for 1050°C (dotted line: end Si, dashed line start SiO$_2$ top layer) Right: Ge counts as a function of depth via SIMS on the same structures.

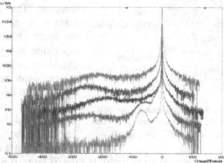

Figure 5: XRD results on the same samples as shown in fig.3. Curves from bottom to top: as grown, 2hrs, 4hrs, 6hrs and 8hrs.

Fig. 5 gives the XRD results for the same samples oxidized at 1050°C. The XRD results are consistent with both the CABOOM and SIMS. XRD confirms the loss of Ge after 8hrs of oxidation.

Analysis using CABOOM and SIMS of the oxidation process at different temperatures shows that for low temperatures the homogeneity of the Ge concentration is poor, but the retention of all Ge atoms in the SiGe layer good (dry oxidation at 1050°C shows the highest Ge concentration peaks within the researched temperature range). For higher temperatures, the peak Ge concentration values are a lot lower due to a loss of Ge before a sufficiently dense top oxide is established, however the homogeneity is improved. For very long oxidation times any process will lead to a loss of Ge. Therefore a two step oxidation process was analysed via CABOOM. The basic idea is to keep a high Ge concentration and create a homogeneous profile. Thus first a low temperature step (1050°C for 2 hrs) is done to retain the Ge in the layer followed by a short high temperature step (1200°C for 1hr) to improve homogeneity without a loss of Ge. The results of the CABOOM characterisation are given in fig.6.

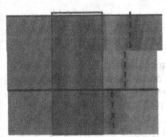

Figure 6: CABOOM results on a two step condensation process. Dashed line marks the estimated bevel edge. The red box surrounds the BOX.

Fig. 6 shows that after an extra hour at 1200°C, approximately 50nm of material is left with a higher Ge concentration at the bevel edge than near the BOX. The blue-green color indicates a Ge concentration between 25 and 35% for our calibrated optical microscope-computer screen system. This layer does seem to be more homogeneous than the sample oxidized for 4 hrs at 1050°C, see fig.4. However, less homogeneous than the 1100°C 5hrs sample presented in fig.3.

CABOOM discussion

Apart from multiple SIMS and TEM, there doesn't exist an easy technique to visualize/determine the influence of the condensation parameters on large areas in one picture. The CABOOM staining technique however is strong in visualizing Ge concentration gradients in a beveled structure. Layer thicknesses and number of layers are very easy to extract without prior layer structure knowledge. Extraction of Ge concentration however relies on the use of numerical tools that fit the modeled spectrum/colors to the measured ones. Another simple, well developed technique is ellipsometry. For ellipsometry, the graded SiGe layer in the Ge condensation samples however proves to be too complex to extract reasonable data. Whilst ellipsometry fails to extract the top oxide thickness that results from the dry oxidation step, CABOOM does give a good estimate notwithstanding the complex graded layer underneath. This is due to the low refractive index of the SiO_2 compared to the SiGe layer underneath. Beveling, to magnify the heterojunction layer width, could also be used for ellipsmetry but in general the width of practical layers remains below the diameter of the ellipsometer laser beam. This is not an issue for the interference pattern in CABOOM. CABOOM however fails when the thickness of the layers are below 30nm.

CABOOM simulations show that the top SiO_2 thickness t_{ox}, of the samples in fig.4 are: t_{ox}=130nm for 2 hrs, t_{ox}=165nm for 4 hrs and t_{ox}=263nm for 8 hrs. These results are consistent with the oxide thickness for dry oxidation at 1050°C on (100) Si surfaces which predicts t_{ox} values of resp. 110, 190 and 280nm. This proves, within experimental error, that indeed the dry oxidation rate for Si and SiGe are the same. It also indicates that the SiGe layers that are being oxidized are virtually stress-free, as strain in the layers is expected to increase the oxidation rate.

CONCLUSIONS

We have shown that CABOOM is an easy technique that gives a fast visual indication on the influence of the chosen condensation process parameters on the Ge concentration for the production of SGOI or GOI. In our experiments we have found that low condensation temperatures are good for the retention of a high Ge contents but poor for homogeneity. This in-homogeneity is not easily removed in subsequent high temperature steps. A long average medium temperature step seems to generate better Ge profiles.

REFERENCES

1. T. Mizuno, S. Takagi, N. Sugiyama, H. Satake, A. Kurobe, and A. Toriumi, "Electron Hole Mobility Enhancement in Strained-Si MOSFETs on SiGe-on-Insulator substrates Fabrication by SIMOX Technology," *IEEE-Elec. Dev. Lett.*, **21**, 230 (2000)
2. Z. Cheng, M. T. Currie, C. W. Leitz, G. Taraschi, E. A. Fitzgerald, J. L. Hoyt, and D. A. Antoniadas, " Electron Mobility Enhancement in Strained-Si n-MOSFETs Fabricated on SiGe-on-Insulator (SGOI) Substrates," *IEEE-Elec. Dev. Lett.*, **22**(7), (2001)
3. T.Tezuka, N.Sugiyama, S.Takagi. Fabrication of strained Si on an ultrathin SiGe-on-insulator virtual substrate with a high-Ge fraction. *App. Phys. Lett.* **79**(12), (2001)
4. N. Hirashita, T. Numata, T. Tezuka,N. Sugiyama, K. Usuda, T. Irisawa,A. Tanabe, Y. Moriyama, S. Nakaharai,S. Takagi, E. Toyoda, Y. Miyamura, "Strained-Si/SiGe-on-Insulator wafers Fabricated by Ge-Condensation Process", 2004 *IEEE Internal. SOI conference*, South Carolina, USA, (Oct 2004)
5. K. Prabhakaran, T. Nishioka, K. Sumitomo, Y. Kobayashi, T. Ogino. Oxidation of ultrathin SiGe layer on Si (001) Evidence for inward movement of Ge". *Jpn. J. Appl. Phys* **33**, 1837 (1994)
6. N. Sugiyama, T. Tezuka, T. Mizuno, M. Suzuki. "Temperature effects on Ge condensation by thermal oxidation of SiGe-on-insulator structures". *J. Appl. Phys.* **95**(8), (2004)
7. K.Kutsukake, N. Usami, K. Fujiwara, T. Ujihara, G. Sazaki, B. Zhang, Y. Segawa, K. Nakajima. "Fabrication of SiGe-On-Insulator througth Thermal Diffusion of Ge on Si-On-Insulator Substrate". *J. Appl. Phys.* **42**, Part 2(3A), (2003)
8. K. Fobelets, T.L. Tan, K. Thielemans, M.M. Ahmad, R.S. Ferguson, and J. Zhang, "Colour coding Ge concentrations in $Si_{1-x}Ge_x$ by bevelling and oxidation: CABOOM", *Semicon Sci Technol* **19**(3),510 (2004)

Mater. Res. Soc. Symp. Proc. Vol. 913 © 2006 Materials Research Society   0913-D01-10

Structure and Process Parameter Optimization for Sub-10nm Gate Length Fully Depleted N-Type SOI MOSFETs by TCAD Modeling and Simulation

Yawei Jin, Lei Ma, Chang Zeng, Krishnanshu Dandu, and Doug William Barlage
ECE, North Carolina State University, Raleigh, NC, 27695

ABSTRACT

According to most recent 2004 International Technology Roadmap for Semiconductor (2004 ITRS), the high performance (HP) MOSFET physical gate length will be scaled to 9nm (22nm technology node) in 2016. We investigate the manufacturability of this sub-10nm gate length fully depleted SOI MOSFET by TCAD simulation. The commercial device simulator ISE TCAD is used. While it is impractical for experiments currently, this study can be used to project performance goals for aggressively scaled devices. In this paper, we will optimize different structure and process parameters at this gate length, such as body thickness, oxide thickness, spacer width, source/drain doping concentration, source/drain doping abruptness, channel doping concentration etc. The sensitivity of device electrical parameters, such as I_{on}, I_{off}, DIBL, Sub-threshold Swing, threshold voltage, trans-conductance etc, to physical variations will be considered. The main objective of this study is to identify the key design issues for sub-10nm gate length Silicon based fully depleted MOSFET at the end of the ITRS. The paper will present the final optimized device structure and optimized performance will be reported.

INTRODUCTION

According to most recent 2004 International Technology Roadmap for Semiconductor (ITRS), the high performance (HP) MOSFET physical gate length will be scaled to 20nm by 2009 (45nm technology node) and 9nm (22nm technology node) by 2016 [1]. The Equivalent physical Oxide Thickness (EOT) will be 0.5nm, the nominal power supply voltage V_{dd} will be only 0.8volts, and the nominal NMOS sub-threshold leakage current at 25°C (I_{sd_leak}) will be about 0.5µA/µm. Device performance increases dramatically due to the decrease of dependence of intrinsic channel resistance. Ng et al. showed the dependence of spreading resistance on lateral abruptness [2] and Taur et al. showed that lateral abruptness affects short channel effects (SCEs) [3]. Kwong et al. discussed the impact of abruptness on a traditional device at 50nm gate length [4]. With shorter gate length, the source/drain doping abruptness and doping concentration have a big impact on resistance which will change device performance dramatically. This paper presents a detailed simulation study of the impact of these parameters as gate length approaches sub-10nm. We choose the fully depleted planar SOI MOSFET at 9nm gate length and compare the SCEs (sub-threshold swing and DIBL), threshold voltage roll-off (for fixed gate work function or fixed I_{off} by choosing proper metal gate work function) and I_{on}-I_{off} characteristics for different source/drain doping abruptness, doping concentration and background doping concentration. The commercial device simulator ISE TCAD is used because it allows precise control of the doping profile. While it is impractical for experiments currently, this study can be used to project performance goals for aggressively scaled devices. The main objective of this paper is to identify the key design issues for sub-10nm gate length Silicon based fully depleted MOSFET at the end of the ITRS.

DEVICE STRUCTURE

The structure of device investigated in this paper is shown in figure 1. The printed gate length is 13nm and the physical gate length L_g is 9nm. In order to get nearly ideal sub-threshold swing and small drain induced barrier lower (DIBL) effect, the silicon body must be very thin and fully depleted. The thickness of silicon body should be approximately 1/3 of gate length for Silicon fully depleted SOI device [5], so a body thickness of 3nm is used.

Oxide thickness has become a major concern as scaling continues. The oxide thickness must be scaled not to improve the speed of the device, but to minimize the short channel effects (SCE) by returning control over the channel to the gate. At thicknesses of around 1.6 nm, SiO2 suffers from an intolerable increase in direct tunneling current [6], i.e. it becomes too leaky. Therefore, high-κ gate dielectrics are being investigated to replace SiO2. Since exact κ values are not known for particular materials, we compare SiO2 and high-κ dielectrics by referring to the equivalent oxide thickness (EOT), which indicates the equivalent thickness of SiO_2 required to produce the same capacitance-voltage curve as obtained from a high-κ dielectric [6]. The oxide thickness is chosen to be 0.5nm according to ITRS2016, in real device at this gate length, high-κ material could replace SiO_2 with the same EOT.

Both industry and research have already headed in the direction of metal gates to mitigate the effects of poly-silicon gate depletion, boron penetration and threshold voltage variation due to dopant fluctuation effects [7]. The major challenges with metal gates are selecting of the proper work function for NMOS and PMOS devices and solving the complex processing issues associated with using metal gates - a discussion which we do not pursue here. Metal is used as the gate electrode in our simulations. Proper gate work function is used to fix I_{off} to ITRS 2016 requirement (0.5μA/μm).

Spacer width is the distance between gate and source/drain, which determines the Miller capacitance of the device. Raised source/drain is used to reduce the series resistance.

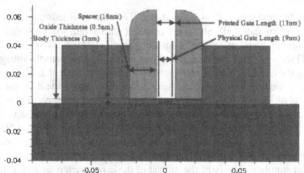

Figure 1. Device Structure for Simulation

DEVICE SIMULATION APPROACH

With 9nm gate length, neither internal nor external characteristics of the MOS device can be described properly using the conventional drift-diffusion transport model. This model cannot reproduce velocity overshoot and often overestimates the impact ionization generation rates. The Monte Carlo method for the solution of the Boltzmann kinetic equation is the most general approach, but because of its high computational requirements, it cannot be used for the routine simulation of devices in an industrial setting. So we use the hydrodynamic (or energy balance) model to describe carrier transport [8] [9].

At this gate length and oxide thickness, quantum mechanical effects are very significant. Density gradient model [10] [11] can include quantization effects by introducing an additional potential quantity in the classical density formula. Because of light doping in the channel, the model for the mobility degradation due to impurity scattering is used. In high electric fields, hydrodynamic Canali model (velocity saturation model) is used [12]. Other physical models included are the recombination model and gate direct tunneling model.

SIMULATION RESULTS

Spacer Width

The width of nitride spacer used in the device is 18nm. The spacer width influences gate capacitance by Miller effect. We change the different spacer width and the gate capacitance is shown in figure 2. It shows that smaller spacer width can lead to higher gate capacitance because of Miller effect. Since the impact of gate-drain capacitance (C_{gd}) is more important than gate-source capacitance (C_{gs}), the asymmetric design may improve this effect. We did not explore asymmetric design in this simulation study but it will be explored in future device optimization. Higher gate capacitance will cause slower switching speed and higher switching power. But if the spacer width is too big, it will increase the series resistance and decrease the drive currents.

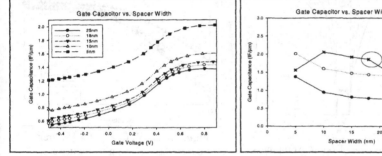

Figure 2. Spacer width Impact on Gate Capacitance

Body Thickness

The Silicon body thickness was chosen as 3 nm because of fully depleted issue [5]. Use of a thicker body will reduce the series resistance and the effect of process variation. But at the same time it will also degrades the short channel effects. We compared the devices with different body thickness in figure 3. Figure 3a shows the I_d-V_g curves with 3nm to 7nm body thickness. With thicker body, the currents increase because of less series resistance, but the device shows substantial off-state leakage current. It is due to the short channel effects. With thicker than 5nm body thickness, the device can not meet ITRS2016 $I_{sd,leak}$ requirement (0.5μA/μm). In figure 3b, we fixed I_{off} to 0.5μA/μm by selecting the proper gate work function and then compare the I_d-V_g curves. 3nm, 4nm and 5nm body thickness show similar I_{on} with same I_{off}. But 3nm body thickness device shows the best sub-threshold swing which leads to a higher turn-on speed and lower switching power. This is because the thinner channel is depleted more easily and controlled by the gate voltage.

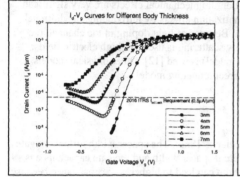 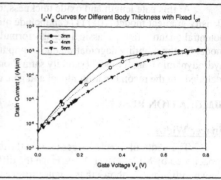

(a) I_d–V_g curves for Different Body Thickness

(b) I_d–V_g curves with Fixed Ioff

Figure 3. Body Thickness Impact

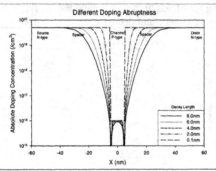 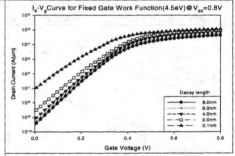

Figure 4. Source/Drain Doping Abruptness

Figure 5. I_d - V_g Curves for Different Source/Drain Doping Adruptness

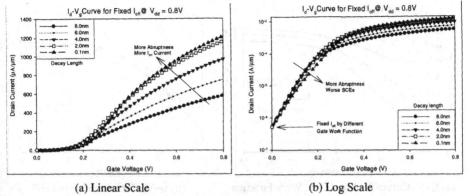

(a) Linear Scale   (b) Log Scale

Figure 6. $I_d - V_g$ Curves for Fixed I_{off}

Doping Abruptness

We change the source/drain lateral doping decay length from 0.1nm to 8nm, which is shown in Fig 4. The source/drain doping is $5*10^{19}/cm^3$ n-type, and background doping is $1*10^{16}/cm^3$ p-type. Fig 5 shows I_d-V_g curves for different S/D doping abruptness. Fig 6 illustrates I_d-V_g curves with fixed I_{off} for each curve by proper gate work function. Fig 7 demonstrates the impact of varying source/drain doping abruptness on device performance. With higher abruptness, I_{on} performance gets better because of less resistance, but SCE gets worse because of the bigger source/drain junction capacitance, which will degrade gate control as well as threshold voltage roll-off.

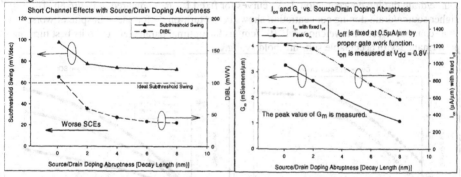

(a) SCEs vs. S/D Doping Abruptness   (b) I_{on}, G_m vs. S/D Doping Abruptness

Figure 7. S/D Doping Abruptness Impacts with Fixed I_{off}

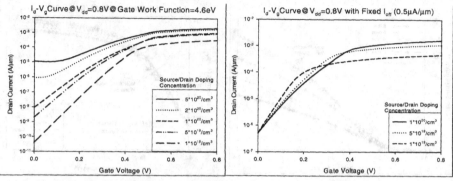

(a) I_d–V_g Curves for Fixed Gate Work Function   (b) I_d–V_g Curves with Fixed Ioff

Figure 8. I_d – V_g Curves for Different Source/Drain Doping Concentration

Source/Drain Doping Concentration

Source/drain doping concentration is another important process parameter for device design. Higher source/drain doping concentration can reduce series resistance which results in good turn-on current performance. On the other hand, higher source/drain doping concentration may lead to bad short channel effect due to higher source/drain junction capacitance. Figure 8 shows the I_d-V_g curves for source/drain doping concentration from $5*10^{20}$/cm^3 to $1*10^{19}$/cm^3. When doping concentration is higher than $1*10^{20}$/cm^3, the leakage current will be higher than ITRS2016 requirements due to GIDL, as shown in figure 8a. So $1*10^{20}$/cm^3 is the limit for source/drain doping concentration. I_d-V_g curves for fixed I_{off} are compared in figure 8b. The higher source/drain doping concentration shows better turn-on performance in spite of worse short channel effects. So we choose $1*10^{20}$/cm^3 as the optimal parameter due to its best turn-on performance with acceptable sub-threshold swing.

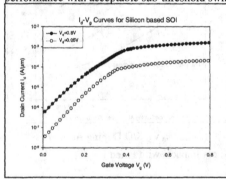 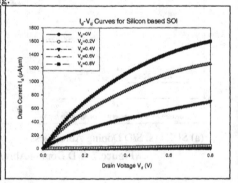

(a) I_d – V_g Curves for Optimized Device   (b) I_d – V_d Curves for Optimized Device

Figure 9. Electrical Performance for Structure and Process Parameter Optimized Device

44

CONCLUSIONS

In summary, we optimized different structure and process parameters at 9nm gate length. The optimized device performance is illustrated in figure 9. To get best I_{on}-I_{off} ratio, we sacrificed some sub-threshold performance. 18nm spacer width and 3nm body thickness are used. Source/drain doping with less than 2nm decay length will have best I_{on}-I_{off} ratio performance. $1*10^{20}/cm^3$ is the limit for source/drain doping concentration because of higher off-state leakage at higher doping concentration. However, the optimized device performance still can't beat ITRS requirements. This indicates that novel structures or materials have to be explored in order to scale below 10 nm and increase device performance further.

REFERENCES

1. ITRS. 2004. Available from URL http://public.itrs.net.
2. K. K. Ng and W. T. Lynch. *IEEE Trans Electron Devices*, 33:965–972, July 1986.
3. Y. taur and D. J. Frank C. H. Wann. *IEDM tec. Dig*, 33:789–792, Dec 1998.
4. M. Kwong, P.Griffin R.Kas and R.Dutton J Plum. *IEEE Trans ED*, 49:1882–1890,2002.
5. R. H. Yuan and K. Lee A. Ourmazd. *IEEE Trans. ED*, 39:1704–1710, 1992.
6. J. S. Suehle. *2001 6th ISPPD*, pages 90–93, May 1997.
7. S.B Samavedam. *IEDM tec. Dig*, pages 247–250, 2002.
8. Y. Apan, B. Polsky, E. Lyumkis and P. Blakey. *IEEE Trans on CAD*, 13:702–710, 1994.
9. S. Szeto and R. Reif. *Solid-State Electronics*, 32(4):307–315, 1989.
10. M. G. Ancona and G. J. Iafrate. *Phys. Rev. B*, 39(13):9536– 9540, May 1989.
11. M. G. Ancona and H. F. Tiersten. *Phys. Rev. B*, 35(15):7959–7965, May 1987.
12. C. Canali and R. Minder G. Majni. *IEEE Trans ED* ED-22:1045–1047, 1975.

Process and Substrate-Induced Strained-Si Development

Mater. Res. Soc. Symp. Proc. Vol. 913 © 2006 Materials Research Society   0913-D02-01

A Novel High-Stress Pre-Metal Dielectric Film to Improve Device Performance for sub-65nm CMOS Manufacturing

Young Way Teh[1], John Sudijono[1], Alok Jain[2], Shankar Venkataraman[2], Sunder Thirupapuliyur[2], and Harry Whitesell[2]

[1]Chartered Semiconductor, Hopewell Junction, NY, 12533
[2]Applied Materials, Santa Clara, CA, 95054

ABSTRACT

This work focuses on the development and physical characteristics of a novel dielectric film for a pre-metal dielectric (PMD) application which induces a significant degree of tensile stress in the channel of a sub-65nm node CMOS structure. The film can be deposited at low temperatures to meet the requirements of NiSi integration while maintaining void-free gap fill and superior film quality such as moisture content and uniformity. A manufacturable and highly reliable oxide film has been demonstrated through both TCAD simulation and real device data, showing ~6% NMOS Ion-Ioff improvement; no Ion-Ioff improvement or degradation on PMOS. A new concept has been proposed to explain the PMD strain effect on device performance improvement. Improvement in Hot Carrier immunity is observed compared to similar existing technologies using high density plasma (HDP) deposition techniques.

INTRODUCTION

Increased doping concentrations required for continued scaling of MOSFET have been shown to result in severe degradation of carrier mobility. At gate lengths of 35nm, channel impurity scattering was projected [1] to dominate mobility. Hence, mobility enhancement using stress becomes a key element in developing next-generation technologies.

HDP is currently the most commonly used gap-fill technology for pre-metal dielectrics (PMD) due to good gap-fill properties, ease of polishing and superior wet etch rate. As geometries shrink, however, the issue of stress and the degradation of NMOS mobility caused by films deposited via HDP technology could be a significant process disadvantage. Scott et. al. [2] first reported a considerable reduction in NMOS Idsat as the device width decreases, for a given device length. They therefore concluded that the effect of HDP compressive stress in STI is more profound on the NMOS device performance; PMOS performance remained unaffected by decreasing width.

In this work, the comparison between HDP oxide and High Aspect Ratio Process (HARP™) oxide film for PMD application is investigated. The strain of both film and its effect on device were characterized and a new concept is proposed to explain the performance improvement phenomenon. HARP is developed using a new O3/TEOS based sub atmospheric chemical vapor deposition process. This film, which is tensile in nature will be used for the PMD layer. This technology is intended to have two key benefits: a greater than 5:1 aspect ratio gap-fill capability, without voids or seams; and the ability to tune an induced stress on the device.

EXPERIMENTAL DETAILS

Shallow Trench Isolation (STI), poly-silicon gate and Nickel silicide layers were formed using a conventional CMOS process. Following these steps, a high tensile silicon nitride film was deposited and selectively etched from the PMOS device and a high compressive silicon nitride layer was then deposited, and selectively etched from an NMOS device. Subsequently, a HARP PMD film was deposited followed by CMP, contact etch and W CMP, as shown in the below flow chart

Process flow for device fabrication

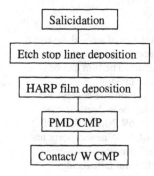

A HARP PMD film with film properties shown in Table 1, was deposited, followed by CMP, contact etch and W CMP. SEM cross-section taken from the NMOS area showed void-free gap fill for aggressive spacing of <13nm, as shown in Figure 1.

Film Parameter	HDP (90/65nm)	HARP-PMD (65/45nm)
Gap filling capability	20nm, 3:1	<10nm, 5:1
Pressure	<10mT	600T
Temperature	400C	400C
Stress	-180 MPa (Comp)	300 MPa (tensile)

Table 1 Film property of HARP film (13nm opening, AR 10:1)

Fig. 1 Void-free gap fill at wafer edge

Electrical performance and explanation

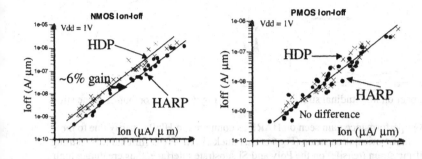

Fig. 2 ~6% Ion-Ioff enhancement on NMOS   Fig. 3 No change in Ion-Ioff on PMOS

Figure 2 showed the NMOS Idsat was improved by 6% with the HARP process as compared to the HDP PMD process, whereas there is no change or degradation in the PMOS performance (Figure 3).

In order to understand the effect of strain on the MOS device, we made reference to early work by C.S. Smith [4] on the piezoresistance effect of semiconductors. The fractional change in silicon resistivity is given by:

$$\frac{\delta\rho}{\rho} = X\,\Pi_{i,j} \text{---(1)}$$

where ρ is the Si resistivity, X is the applied stress and $\Pi_{i,j}$ is the piezoresistance coefficient for different crystallographic orientations i,j

The change in Si resistance is strongly dependent on the piezoresistance coefficient. For a [001] Si surface and notch along <110> wafer, the calculated piezoresistance coefficients for both N and P type Si with respect to stress orientation (Fig 4) are listed in Table 2. A positive coefficient with a corresponding compressive strain will produce lower Si resistivity for P type; and a negative coefficient with a corresponding tensile strain will produce lower Si resistivity for N type. The larger the coefficient, the more sensitive is the pizoresistance effect to the stress.

Current flow	Stress orientation	Piezo Coefficient (10-12 cm2/dyne)	
		N Type Si	P type Si
Ex	Ex	-31.2	+ 71.8
Ex	Ey	-17.6	+ 66.3
Ex	Ez	+ 53.4	-1.1

Fig 4. Stress orientation in Transistor   Table 2 : Piezoresistance coefficient for different stress orientation

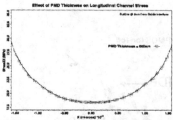
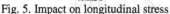
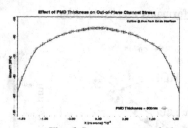

Fig. 5. Impact on longitudinal stress   Fig. 6. Impact on out-of-plane stress

The NMOS performance gain seen on HARP as compared to HDP is partly due to the strong tensile Ez orientation strain (Fig 6). This thick HARP PMD film induces a strong pulling/lifting strain (tensile) on the Poly and Si substrate interface, thus creating a high compressive residue strain on the Si channel in the Ez orientation. This compressive strain in the channel will significantly boost the NMOS performance due to the high piezoresistance coefficient (+53.4). At the same time, this PMD will also induce a compressive strain in the PMOS. But the Piezoresistance coefficient shows that the PMOS is insensitive to the Ez strain effect (-1.1). Therefore, the PMOS performance remains unchanged.

Conclusions:
A novel dielectric HARP film has been developed using a new O_3/TEOS based fill process. This technology is intended to have two key benefits: a void-free gap fill on greater than 5:1 aspect ratio features, and the ability to tune an induced stress to enhance device performance. A manufacturable and highly reliable oxide film has been demonstrated on device data with a ~6% NMOS Ion-Ioff improvement, and no degradation on PMOS. Improvement in Hot Carrier performance is seen, as compared to similar existing technologies using HDP techniques. Combined with a tensile pre-metal dielectric (PMD), stress management and optimization of the above films can yield significant performance improvements without additional cost, or integration complexities.

References:
1. T.Ghani et al., Symp. VLSI Tech. Dig., pp.174-175, 2000.
2. G. Scott, et al., IEDM pp. 827-830 (1999).
3. S. Ito et al., IEDM 2000, p. 247.
4. C. S. Smith, Piezoresistance affect in Germanium and Silicon, Phys. Rev. , Vol. 94 ; pp42-49, 1954

Mater. Res. Soc. Symp. Proc. Vol. 913 © 2006 Materials Research Society   0913-D02-02

Mobility Enhancement by Strained Nitride Liners for 65nm CMOS Logic Design Features

Claude Ortolland[1,2], Pierre Morin[3], Franck Arnaud[3], Stephane Orain[1], Chandra Reddy[4], Catherine Chaton[5], and Peter Stolk[1]

[1]Philips Semiconductors, 860, rue Jean Monnet, Crolles, Isere, 38926, France
[2]Laboratoire Physique de la Matirère, 7, Avenue Jean Capelle, Villeurbanne, Rhone, 69621, France
[3]ST Microelectronics, 850, rue Jean Monnet, Crolles, Isere, 38926, France
[4]Freescale Semiconductor, 870, rue Jean Monnet, Crolles, Isere, 38926, France
[5]CEA-LETI, 850, rue Jean Monnet, Crolles, Isere, 38926, France

ABSTRACT

In this paper the impact of process-induced stress and transistor layout on device performance in state-of-the-art 65nm CMOS technology has been studied. We have focused this analysis on different nitride liners above devices (Contact Etch-Stop Layers – CESL) which have been fabricated on two differently oriented (100) substrates: <110> and <100>. This overview permits to have a good understanding of CESL, and to choose the right strategy in terms of process induced stress in future microelectronic technologies.

INTRODUCTION

Recent publications have shown the advantages of CESL [1] techniques to boost intrinsic MOS transistors performance. It is well known that performance enhancement is achieved on nMOS with tensile film whereas on pMOS with high compressive layers [2]. Moreover, substrate orientation engineering proves that <100> channel enhances pMOS drive current without significant effect on nMOS [3] & [4]. It has been demonstrated that thickness, intrinsic stress and Sidewall Step Coverage (SSC) are important material parameters to reach the highest mobility enhancement for different devices [5]. In this paper, an analysis of CESL and substrate orientation on device properties is presented. A specific attention has been paid to relate the performance gain to the active layout and transistor dimension. In addition CESL nitride layers grown by Plasma Enhance Chemical Vapour Deposition (PECVD) and Atomic Layer Deposition (ALD) are compared.

EXPERIMENTAL DETAILS

A conventional 65nm process flow has been integrated on (100) bulk substrates with a $0.3\mu m$ STI depth and with either <100> or <110> current flow orientation. A 1.8nm oxynitride and 100nm of poly thickness compose the Gate stack. NiSi is used as salicide. The purpose is to investigate the layout parameters which modulate the CESL effect on transistor performance. The set of nitride experiments is described by table I and II. Different nitride processes for Tensile CESL have been compared to determine the key material parameters.

Table I and II. Different nitrides used for CESL with PECVD Process (I: left table) and ALD process (II: right table).

Sample PECVD	Deposition Temperature [°C]	Thickness [nm]	Stress As Dep. [GPa]
1	400	33	1,1
2	400	35	1,2
3	400	47	1,2
4	400	58	1,2
5	400	81	1,2
6	400	35	0,5
7	400	35	0,9
8	400	50	0,9
9	480	35	0
10	480	50	0,75
11	480	35	0,75
12	480	120	0,75
13	480	35	1,2
14	480	50	1,2
15	400	50	1,5
16	400	50	-1,9

Sample ALD	Deposition Temperature [°C]	Thickness [nm]	Stress As Dep. [GPa]
A	350	38	1,37
B	350	55	1,26
C	390	54	1,47
D	400	37	0,86
E	450	36	0,95
F	450	77	1,17

In parallel we have studied the impact of substrate orientation as described in table III. The purpose is to investigate the correlation between channel orientation and CESL stress. The substrate surface plane is (100).

Table III. Stressed CESL with substrate splits.

Channel orientation	Contact Etch Stop Layer	Comment
(100) <100>	Control	Sample 6
	Tensile	Sample 14
	Compressive	Sample 16
(100) <110>	Control	Sample 6
	Tensile	Sample 14
	Compressive	Sample 16

INFLUENCE OF TRANSISTOR LAYOUT

The device performances obtained with strained CESL are very sensitive to gate width, gate length and gate pitch. For nMOS, the channel area (W x L) is found to be the prime gain modulator (Figure 1). Mechanical simulation has testified that the main CESL strain effect stands in the vertical direction. This is consistent with a linear relationship between the drive current gain and the transistor area observed in Figure 1.

The dimensional scaling observed with each technology node is linked with an increase of the transistor density. This equivalent to a decrease of the distance between poly gates (poly pitch). The impact of the decrease of this poly pitch is depicted in Figure 2 & Figure 3.

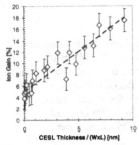

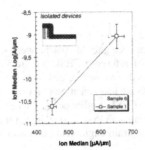

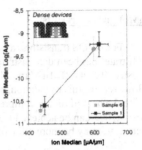

Figure 1. nMOS Ion gain versus CESL thickness divided by gate area surface.

Figure 2. Ioff-Ion median with isolated nMOS devices

Figure 3. Ioff-Ion median with dense nMOS devices

Ion-Ioff merit curves between nMOS with low or high CESL strain show less stress induced gain with dense pitch (Figure 3) than in the case of isolated transistors (Figure 2). This phenomenon has been clearly explained by simulation in [6].

INFLUENCE OF SUBSTRATE ORIENTATION

nMOS device results

The comparison of electrical Ioff-Ion performance (Figure 4), either with tensile or compressive liners, shows exactly the same behaviour whatever the substrate orientation. Tensile (compressive) CESL boosts (degrades) the devices similarly for the two orientations. Similar gain as a function of gate length are observed in Figure 5 whatever channel orientation, showing that strain modulation for nMOS can be universally applied.

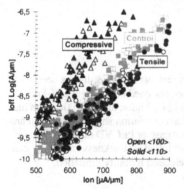

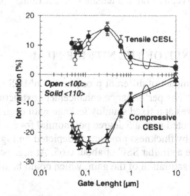

Figure 4. nMOS Ioff-Ion merit curve on two different substrates with compressive, neutral (control) and tensile Contact Etch Stop Layer. W=1μm

Figure 5. Ion variation due to strained CESL versus gate length for the two different substrates. W=1μm

pMOS device results

For pMOS transistors, a performance gain of 10% between <100> and <110> oriented substrates has been demonstrated with reference CESL (Figure 6 & Figure 7: -335 mA/μm versus -305 @Ioff=1nA/μm). This is due to the reduced mass of heavy holes in <100> direction. However, for this orientation, pMOS are less sensitive to the strain induced within the channel. With the help of highly compressive CESL, higher performance on <110> oriented substrates than <100> can be achieved (-360 versus -345 mA/μm), indicating that <110> oriented substrates may be more attractive for strain engineering on future CMOS technologies.

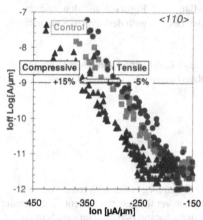 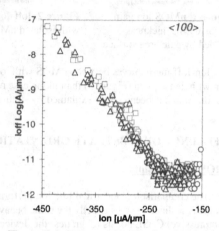

Figure 6. Ioff-Ion merit curve for pMOS device on <110> substrate orientation with compressive, neutral and tensile Contact Etch Stop Layer.

Figure 7. Ioff-Ion merit curve for pMOS device on <100> substrate orientation with compressive, neutral and tensile Contact Etch Stop Layer.

INFLUENCE OF THE NITRIDE FILM

The impact of different processes on nMOS <110> parameters (Table I & II) has been analysed. The performance enhancement extraction has been obtained on different Ioff-Ion merit curves at a fixed standby leakage with each process (Figure 8): Ion at Ioff=100nA/μm has been extracted with a 5 degree polynomial for each split. A linear relation ship between gain and the stress by thickness products is depicted on Figure 9 in the case of PECVD nitride. All these films have a similar SSC of about 0.65. At highest stress multiplied by thickness values, we can observe a saturation of the gain, which could be due to the electron mobility limitation.

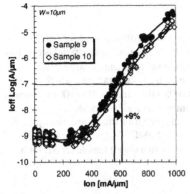

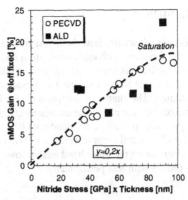

Figure 8. Ioff-Ion merit curve with 2 different CESL process

Figure 9. Ion nMOS gain at Ioff fixed versus nitride stress multiplied by thickness for different CESL processes.

Contrary to PECVD nitride, the ALD process presents a very good Sidewall Step Coverage, at about 0.95. As a consequence, this kind of layer presents a higher drive current enhancement than PECVD films when it is processed at high temperature (450°C) [5]. If we compare it at equivalent force (Stress x Thickness) a lower gain induced by PECVD nitride is obtained as expected.

Figure 11 shows the nMOS Ion gain due to ALD liner as a function of the gate Length. Two typical behaviours are observed. First for the long device with L>100nm, the impact of the thickness and the intrinsic stress of ALD liner is visible. Figure 12 plots the normalized gain divided by the product of stress times thickness, for a 0.25μm gate length. A clear dependence of this gain versus deposited temperature is visible. This mean that 1MPa of ALD nitride deposited at 350°C brings 6 times less performance improved than 1MPa at higher nitride temperature.

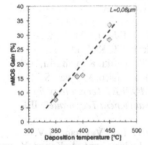

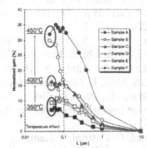

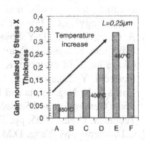

Figure 10. nMOS Ion gain for the nominal transistor (L=60nm) & intrinsic stress of the ALD films versus the deposition temperature.

Figure 11. nMOS Ion gain versus gate length for the different ALD CESL.

Figure 12. Normalized Ion nMOS gain with gate length = 0.25μm.

CONCLUSION

This paper describes how the transistor behaviour responds to different kinds of CESL stresses as a function of channel orientation and device layout. It is found that material properties can have significant impact on strain transmission into the conduction channel: stress level, thickness, Sidewall Step Coverage and deposition temperature are some examples which have been presented. The channel orientation choice between <100> (100) and <110> (100) substrates has no importance for nMOS strain sensibility. But it becomes a critical parameter for a technology due to pMOS performance: <110> oriented substrates with a dual stress liner process (tensile for nMOS and compressive for pMOS) enable a boosted pMOS but at higher cost [7]. <100> substrates on the other hand enable lower cost process with a single tensile CESL for a standard pMOS with less performance.

REFERENCES

1. S. Ito, H. Namba, K. Yamaguchi, T. Hirata, K. Ando, S. Koyama, S. Kuroki, N. Ikezawa, T. Suzuki, T. Saitoh and T. Horiuchi, "*Mechanical Stress Effect of Etch-Stop Nitride and its Impact on Deep Submicron Transistor Design*", IEDM Technical Digest 2000, pp. 247-250
2. C. Ortolland, S. Orain, J. Rosa, P. Morin, F. Arnaud, M. Woo, A. Poncet and P. Stolk, "*Electrical Characterization and Mechanical Modeling of Process Induced Strain in 65 nm CMOS Technology*", ESSDERC Proceeding 2004, pp. 137-140
3. H. Sayama, Y. Nishida, H. Oda, T. Oishi, S. Shimizu, T. Kunikiyo, K. Sonoda, Y. Inoue and M. Inuishi, "*Effect of <100> Channel Direction for High Performance SCE Immune pMOSFET with Less Than 0.15⎕m Gate Length*", IEDM Technical Digest 2004, pp. 657-660
4. T. Komoda, A. Oishi, T. Sanuki, K. Kasai, K. Yoshimura, K. Ohno, M. Iwai, M. Saito, F. Matsuoka, N. Nagashima and T. Noguchi "*Mobility Improvement for 45nm Node by Combination of Optimized Stress Control and Channel Orientation Design*", IEDM Technical Digest 2004, pp. 217-220
5. P. Morin, C. Chaton, C. Reddy, C. Ortolland, M.T. Basso, F. Arnaud and K. Barla, "*Tensile Contact Etch Stop silicon nitride for nMOS performance enhancement: influence of the film morphology*", ECS Proceeding 2005, pp. 253-263
6. A. Oishi, O. Fujii, T. Yokoyama, K. Ota, T. Sanuki, H. Inokuma, K. Eda, T. Idaka, H. Miyajima, S. Iwasa, H. Yamasaki, K. Oouchi, K. Matsuo, H. Nagano, T. Komoda, Y. Okayama, T. Matsumoto, K. Fukasaku, T. Shimizu, K. Miyano, T. Suzuki, K. Yahashi, A. Horiuchi, Y. Takegawa, K. Saki, S. Mori, K. Ohno, I.Mizushima, M. Saito, M. Iwai, S. Yamada, N. Nagashima and F. Matsuoka,"*High Performance CMOSFET Technology for 45nm Generation and Scalability of Stress-Induced Mobility Enhancement Technique*", IEDM Technical Digest 2005, pp. 239-242
7. C.D. Sheraw, M. Yang, D.M. Fried, G. Costrini, T. Kanarsky, W-H. Lee, V. Chan, M.V. Fischetti, J. Holt, L. Black, M. Naeem, S. Panda,L. Economikos, J. Groschopf, A. Kapur, Y. Li, R. T. Mo, A. Bonnoit, D. Degraw, S. Luning, D. Chidambarrao, X. Wang, A. Bryant, D. Brown,C-Y. Sung, P. Agnello, M. Ieong, S-F. Huang, X. Chen, M. Khare "*Dual Stress Liner Enhancement in Hybrid Orientation Technology*", VLSI Technology symposium 2005, pp.12-13

Mater. Res. Soc. Symp. Proc. Vol. 913 © 2006 Materials Research Society   0913-D02-04

Process-Induced Strained P-MOSFET Featuring Nickel-Platinum Silicided Source/Drain

Rinus Tek Po Lee[1], Tsung-Yang Liow[1], Kian-Ming Tan[1], Kah-Wee Ang[1], King-Jien Chui[1], Qiang-Lo Guo[2], Ganesh Samudra[1], Dong-Zhi Chi[3], and Yee-Chia Yeo[1]

[1]Electrical and Computer Engineering, National University of Singapore, Silicon Nano Device Laboratory, Singapore, Singapore, 119260, Singapore
[2]Institute of Microelectronics, 11 Science Park Road, Science Park-II, Singapore, Singapore, 117685, Singapore
[3]Institute of Materials Research and Engineering, 3 Research Link, Singapore, Singapore, 117602, Singapore

ABSTRACT

We report the use of nickel-platinum silicide (NiPtSi) as a source/drain (S/D) material for strain engineering in P-MOSFETs to improve drive current performance. The material and electrical characteristics of NiPtSi with various Pt concentrations was investigated and compared with those of NiSi. $Ni_{0.95}Pt_{0.05}Si$ was selected for device integration. A 0.18 μm gate length P-MOSFET achieved a 22% gain in I_{Dsat} when $Ni_{0.95}Pt_{0.05}Si$ S/D is employed instead of NiSi S/D. The enhancement is attributed to strain modification effects related to the nickel-platinum silicidation process.

INTRODUCTION

Strained silicon technology has recently been actively explored for the improvement of drive current I_{Dsat} and carrier mobility in metal-oxide-semiconductor field-effect transistors (MOSFETs) [1-5]. One of the most cost-effective and manufacturable options of introducing strain in the transistor channel is through the exploitation of process-induced strain. Process-induced strain can be introduced at various process steps during transistor fabrication, e.g. metal silicidation of source/drain (S/D), shallow-trench isolation, and contact-etch-stop layer [2-4]. With the reduction of the transistor gate length L_G, strain induced by the metal silicide in the S/D regions becomes increasingly significant. Engineering the strain due to the silicided S/D could be an attractive approach to further enhance the I_{Dsat} performance. Recently, both N- and P-MOSFETs have been reported to exhibit improved transconductance G_m and I_{Dsat} resulting from silicide-induced strain in the S/D region [2, 5]. Currently, studies on silicide-induced strain focus on cobalt silicide ($CoSi_2$) and nickel silicide (NiSi). The origin of this strain is attributed to the reduction in volume during silicidation, the mismatch of the thermal expansion coefficients of silicide and silicon, or the lattice mismatch between the silicide and silicon. Volume reduction during silicidation and the mismatch in thermal expansion coefficients of silicide and silicon play an important role in the introduction of strain in the transistor channel (Figure 1). Modification of the induced strain through silicide materials engineering could be explored for device performance enhancement. In this paper, we report a novel approach for improving the drive current performance in P-MOSFETs by employing nickel-platinum as the S/D silicide material. The material characteristics of nickel-platinum silicide (NiPtSi) with different Pt concentrations will be discussed, and an optimum Pt concentration to improve P-MOSFET performance is demonstrated.

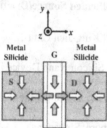

(a) Strain Due to Metal Silicide Formation

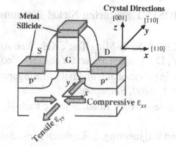

(b) Strain Components for Enhancement of P-channel Transistor Drive Current

- N-well and V_T Adjust Implants
- SiO$_2$ Gate Oxidation (EOT = 4 nm)
- Poly-Silicon Gate Deposition
- Gate Lithography, Trimming, Etch
- SiN/SiO$_2$ Spacer Formation
- S/D Implant and Activation @ 1000 °C, 1 s
- Ni or NiPt Silicidation @ 500 °C, 60 s
- Selective Metal Etch

Figure 1. (a) The larger thermal expansion coefficient of metal silicide compared with Si as well as the volume reduction during silicidation contribute to tensile strain ε_{xx} and compressive strain ε_{yy} in the transistor channel. These strain components degrade hole mobility. (b) For improved performance in P-MOSFET, the channel strain components are preferably compressive (or less tensile) in the [110] channel direction and tensile (or less compressive) in the perpendicular y-direction.

Figure 2. Process sequence employed in the fabrication of P-MOSFETs with nickel (Ni) and nickel-platinum (NiPt) silicided source/drain regions.

EXPERIMENTAL DETAILS

P-MOSFETs with S/D metal silicide comprising NiSi or NiPtSi with various Pt concentrations were fabricated. The device fabrication process sequence is shown in Figure 2. (001) Si substrates were used and all fabricated transistors have the <110> channel direction. Well and threshold voltage adjust implants were performed followed by thermal oxidation to form SiO$_2$ gate dielectric (EOT ~ 4 nm). Boron-doped poly-silicon was employed as the gate electrode. 248 nm optical lithography and photoresist trimming were performed to obtain linewidths down to 0.18 μm. This was followed by poly-silicon gate etch and source/drain extension ion implantation. Silicon nitride (SiN) spacers with SiO$_2$ liner were formed. The spacer width was approximately 25 nm. S/D ion implantation and dopant activation at 1000°C for 1 second using rapid thermal annealing (RTA) were subsequently performed. Thin films of nickel and nickel-platinum were deposited by e-beam evaporation. Silicidation of the S/D regions employed annealing at 500°C for 60 s in a N$_2$ ambient. Selective metal etch was then performed with a SPM solution [H$_2$SO$_4$:H$_2$O$_2$ (4:1)] at 70°C for 60 s to remove any unreacted metal. The metal thickness deposited for the nickel silicided devices and nickel-platinum silicided devices were such that the final Ni- and NiPt- silicide thicknesses in the S/D regions are similar (~40 nm). This completed the device fabrication and the devices were directly probed.

Material analysis was also performed on blanket wafers. Ni and NiPt were deposited and subjected to a 60 s rapid thermal anneal (RTA) in N$_2$ ambient at temperatures ranging from 300 to 600°C. X-ray diffraction (XRD) and four-point probe (FPP) measurements were employed in the analysis of the metal silicide layers.

RESULTS AND DISCUSSION

XRD analysis and FPP measurements were carried out to investigate the impact of different Pt concentrations on the material characteristics of NiPtSi. This allowed the selection of the optimum Pt concentration in NiPtSi for subsequent device integration. Two different Pt concentrations in NiPtSi were investigated for this work. The compositions of the NiPtSi films were determined by Rutherford Backscattering Spectroscopy (RBS) to be $Ni_{0.95}Pt_{0.05}Si$ and $Ni_{0.88}Pt_{0.12}Si$. Figures 3, 4, and 5 show the glancing-angle ($\theta/2\theta$) XRD scans of NiSi, $Ni_{0.95}Pt_{0.05}Si$, and $Ni_{0.88}Pt_{0.12}Si$ thin films as a function of silicidation temperatures ranging from 300 to 600 °C. The XRD peaks are labeled according to the MnP structure of NiSi [6]. We observed that the peak positions generally shift toward lower 2θ angles with increasing Pt concentration, indicating the expansion of the NiSi lattice due to the incorporation of Pt. It is also observed that increasing the Pt concentration changes the relative peak intensities appreciably, suggesting preferential crystal orientation due to the incorporation of Pt. This is possibly due to the epitaxial alignment of the NiSi (020) or NiSi (013) planes parallel to the (001) Si substrate [7].

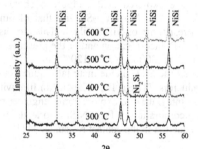

Figure 3. XRD scans of NiSi films as a function of silicidation temperatures ranging from 300–600 °C. The XRD peaks are labeled according to the MnP structure of NiSi.

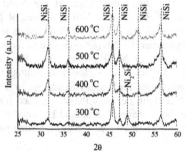

Figure 4. XRD scans of $Ni_{0.95}Pt_{0.05}Si$ films as a function of silicidation temperatures ranging from 300–600 °C. Noticeable shift of NiSi peaks to lower 2θ angles is observed.

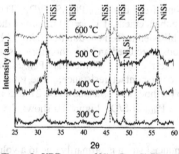

Figure 5. XRD scans of $Ni_{0.88}Pt_{0.12}Si$ films as a function of silicidation temperatures ranging. Significant shift of NiSi peaks to lower 2θ angles is observed.

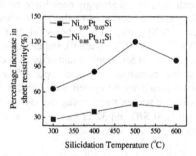

Figure 6. Percentage change in sheet resistivity for $Ni_{0.95}Pt_{0.05}Si$, and $Ni_{0.88}Pt_{0.12}Si$ films as a function of silicidation temperatures, compared with the sheet resistivity of NiSi films.

For pure Ni siliciation, it is known that the metal-rich high-resistivity di-nickel silicide (Ni_2Si) phases are present only from temperatures between 250 to 350°C for pure Ni-Si silicidation reactions [8]. The addition of Pt has been shown to elevate the phase transformation temperature of Ni_2Si-to-NiSi considerably [9]. Therefore, it is important to consider the Ni_2Si-to-NiSi phase transformation temperature in the selection of an optimum Pt concentration in NiPtSi. Careful examination of the XRD plot in Figure 5 indicates that the Ni_2Si and NiSi phases co-exist in the $Ni_{0.88}Pt_{0.12}Si$ film following silicidation at 300°C, and possibly up to 500°C. It is believed that the Ni_2Si phase is present in $Ni_{0.88}Pt_{0.12}Si$ film for anneal temperatures of up to 500°C. In contrast, the Ni_2Si phase is clearly absent in the $Ni_{0.95}Pt_{0.05}Si$ film annealed at 500°C (Figure 4). The Ni_2Si-to-NiSi phase transformation temperature for $Ni_{0.95}Pt_{0.05}Si$ is estimated to be about 400°C.

Transformation into the low resistivity NiSi phase is important for achieving low series resistance in the S/D regions for good device performance. The total S/D series resistance comprises several components, including that due to the resistivity of the metal silicide, the contact resistance between the silicide material and silicon, and the resistivity of the doped S/D regions. The sheet resistivities of the NiPtSi films are explored next. Figure 6 shows a comparison of the sheet resistivity values of $Ni_{0.95}Pt_{0.05}Si$ and $Ni_{0.88}Pt_{0.12}Si$ with that of pure NiSi. In general, the addition of Pt increases the sheet resistivity of NiPtSi films. This increase is more significant for $Ni_{0.88}Pt_{0.12}Si$ due to the significant Pt incorporation and/or the presence of the high-resistivity Ni_2Si phase. $Ni_{0.88}Pt_{0.12}Si$ formed at 500°C has a 120% higher sheet resistivity than NiSi, and is clearly inadequate in meeting the S/D silicide material requirements in the International Technology Roadmap for Semiconductors (ITRS) [10]. The NiPtSi film with a lower Pt content, i.e. $Ni_{0.95}Pt_{0.05}Si$, is used for device integration to explore strain engineering in the transistor channel. It should be noted that the incorporation of Pt in NiSi also leads to a reduction in the hole barrier energy, contributing to a reduction in the contact resistance [11].

The I_{DS}-V_{DS} characteristics of a 0.18 μm gate length P-MOSFET with $Ni_{0.95}Pt_{0.05}Si$ S/D shows an enhancement of 22% in I_{Dsat} compared with a control P-MOSFET with NiSi S/D at a gate over-drive (V_{GS} - V_t) of 1.0 V (Figure 7). The I_{Dsat} enhancement as a function of L_G for P-MOSFET with $Ni_{0.95}Pt_{0.05}Si$ S/D compared with P-MOSFET with NiSi S/D is shown in Figure 8. The I_{DS} enhancement is strongly dependent on the L_G and increases dramatically with reduced L_G. This could be indicative of the strain effect arising from the $Ni_{0.95}Pt_{0.05}Si$ S/D. The P-MOSFET with $Ni_{0.95}Pt_{0.05}Si$ S/D shows substantial G_m enhancement at V_{DS} = 50 mV over the P-MOSFET with NiSi S/D (Figure 9), indicating hole mobility enhancement due to strain. In both transistors, identical S/D implant and activation conditions were used and it is therefore assumed that the series resistance due to the doped S/D regions are the same. While $Ni_{0.95}Pt_{0.05}Si$ has a higher sheet resistivity compared with NiSi, it has a lower contact resistance. The total S/D series resistance was extracted using the shift-and-ratio method, and obtained to be 27 Ω for the device with NiSi S/D and 32 Ω for the device with $Ni_{0.95}Pt_{0.05}Si$ S/D. Figure 10 shows the I_{DS}-V_{GS} characteristics for both devices.

Recent reports have demonstrated that silicide-induced strain is primarily due to a volume reduction during the silicidation process and the mismatch of thermal expansion coefficient between the metal silicide and Si [12]. As illustrated in Figure 1, the contribution of silicide-induced stress to the in-plane strain components in the channel region is believed to be tensile in the source-to-drain channel direction, i.e. [110], and compressive in the perpendicular direction. NiSi and PtSi each exhibits volume reduction of ~ 22% and 11%, respectively, during silicidation, and the thermal expansion coefficient of NiSi and PtSi are 16×10^{-6} /K and 14×10^{-6}/K, respectively [12, 13]. The addition of Pt in NiSi should reduce the amount volume contraction during silicidation, and also reduce the strain due to thermal expansion coefficient mismatch with Si. By using NiPtSi as the S/D silicide, it is expected that the tensile ε_{xx} and compressive ε_{yy} strain components in the transistor channel would be reduced compared to NiSi. As a result, hole mobility and drive current are improved in P-MOSFETs employing $Ni_{0.95}Pt_{0.05}Si$ S/D (Figure 7 and Figure 9).

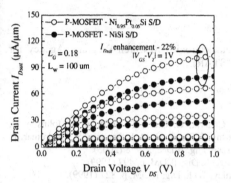

Figure. 7. I_{DS}-V_{DS} characteristics for P-MOSFET with NiSi S/D and P-MOSFET with $Ni_{0.95}Pt_{0.05}Si$ S/D at various gate over-drives.

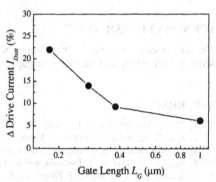

Figure 8. I_{Dsat} enhancement for P-MOSFET with $Ni_{0.95}Pt_{0.05}Si$ S/D as a function of L_G @ $|V_{GS} - V_T| = |V_{DS}| = 1V$.

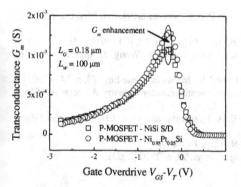

Figure 9. Comparison of maximum transconductance, G_m at the same gate over-drive of P-MOSFET with NiSi S/D and P-MOSFET with $Ni_{0.95}Pt_{0.05}Si$ S/D.

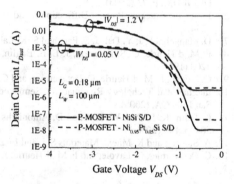

Figure 10. I_{DS}-V_{GS} characteristics for P-MOSFETs with NiSi S/D and $Ni_{0.95}Pt_{0.05}Si$ S/D.

CONCLUSION

We explored the material and electrical characteristics of NiSi and NiPtSi with various Pt concentrations formed at temperatures ranging from 300 to 600°C. Transistors with Ni silicided S/D and NiPt silicide S/D were fabricated. A 0.18 μm gate length P-MOSFET with $Ni_{0.95}Pt_{0.05}Si$ S/D delivered I_{Dsat} enhancement of 22% when compared with a P-MOSFET with NiSi S/D. The enhancement is attributed to an overall decrease of tensile ε_{xx} and compressive ε_{yy} strain components in the transistor channel with the incorporation of Pt. The nickel-platinum silicidation process is a simple and viable technique to selectively optimize the drive current of P-MOSFETs. Materials engineering focusing on the adjustment of silicide-induced strain effects could be explored as potential performance-boosters for advanced CMOS devices.

ACKNOWLEDGEMENTS

We acknowledge a research grant from the Nanoelectronics Research Program, Agency for Science, Technology and Research, Singapore (A*STAR).

REFERENCES

1. T. Ghani, M. Armstrong, C. Auth, M. Bost, P. Charvat, G. Glass, T. Hoffman, K. Johnson, C. Kenyon, J. Klaus, B. McIntyre, K. Mistry, A. Murthy, J. Sandford, M. Silberstein, S. Sivakumar, P. Smith, K. Zawadzki, S. Thompson, and M. Bohr, Tech. Dig. IEDM, p. 978 (2003).
2. A. Steegen, M. Stucchi, A. Lauwers, and K. Maex, Tech. Dig. IEDM, p. 497 (1999).
3. G. Scott, J. Lutze, M. Rubin, F. Nouri, and M. Manley, Tech. Dig. IEDM, p. 827 (1999).
4. S. Ito, H. Namba, K. Yamaguchi, T. Hirata, K. Ando, S. Koyama, S. Kuroki, N. Ikezawa, T. Suzuki, T. Saitoh, and T. Horiuchi, Tech. Dig. IEDM, p. 247 (2000).
5. C. H. Ge, C. C. Lin, C. H. Ko, C. C. Huang, Y. C. Huang, B. W. Chan, B. C. Perng, C. C. Sheu, P. Y. Tsai, L. G. Yao, C. L. Wu, T. L. Lee, C. J. Chen, C. T. Wang, S. C. Lin, Y.-C. Yeo, and C. Hu, Tech. Dig. IEDM, p. 73 (2003).
6. Joint Committee on Powder Diffraction Standards (J.C.P.D.S), powder diffraction files, Card No. 38-0844.
7. D. Mangelinck, J. Y. Dai, J. S. Pan, and S. K. Lahiri, Appl. Phys. Lett. 73, p. 1736 (1999).
8. F. M. d'Heurle, C. S. Peterson, J. E. E. Baglin, S. Laplaca, and C. Y. Wong, J. Appl. Phys. 55, p. 4208 (1984).
9. C. Lavoie, F. M. d'Heurle, C. Detavernier, and C. Cabral Jr., Microelectronic Eng. 70, p. 144 (2003).
10. International Techology Roadmap for Semiconductors, Semiconductor Industry Association (SIA), San Jose, CA, (2005).
11. L. J. Jin, K. L. Pey, W. K. Choi, D. Antoniadis, E. A. Fitzgerald, and D. Z. Chi, Thin Solid Films, 504, p. 149 (2006).
12. A. Steegen and K. Maex., Materials Sci. and Eng. R. 38, p. 1 (2002).
13. C. Detavernier, C. Lavoie, and F. M. d'Heurle, J. Appl. Phys. 93, p. 2510 (2003).

Mater. Res. Soc. Symp. Proc. Vol. 913 © 2006 Materials Research Society 0913-D02-05

Thermal Stability of Thin Virtual Substrates for High Performance Devices

Sarah H Olsen[1], Steve J Bull[1], Peter Dobrosz[1], Enrique Escobedo-Cousin[1], Rimoon Agaiby[1], Anthony G O'Neill[1], Howard Coulson[2], Cor Claeys[3], Roger Loo[3], Romain Delhougne[3], and Matty Caymax[3]

[1]University of Newcastle, Newcastle upon Tyne, NE1 7RU, United Kingdom
[2]Atmel North Tyneside, Newcastle upon Tyne, NE28 9NZ, United Kingdom
[3]IMEC, Leuven, B-3001, Belgium

ABSTRACT

Detailed investigations of strain generation and relaxation in Si films grown on thin $Si_{0.78}Ge_{0.22}$ virtual substrates using Raman spectroscopy are presented. Good virtual substrate relaxation (>90%) is achieved by incorporating C during the initial growth stage. The robustness of the strained layers to relaxation is studied following high temperature rapid thermal annealing typical of CMOS processing (800-1050 °C). The impact of strained layer thickness on thermal stability is also investigated. Strain in layers below the critical thickness did not relax following any thermal treatments. However for layers above the critical thickness the annealing temperature at which the onset of strain relaxation occurred appeared to decrease with increasing layer thickness. Strain in Si layers grown on thin and thick virtual substrates having identical Ge composition and epilayer thickness has been compared. Relaxation through the introduction of defects has been assessed through preferential defect etching in order to verify the trends observed. Raman signals have been analysed by calibrated deconvolution and curve-fitting of the spectra peaks. Raman spectroscopy has also been used to study epitaxial layer thickness and the impact of Ge out-diffusion during processing. Improved device performance and reduced self-heating effects are demonstrated in thin virtual substrate devices when fabricated using strained layers below the critical thickness. The results suggest that thin virtual substrates offer great promise for enhancing the performance of a wide range of strained Si devices.

INTRODUCTION

Novel device structures and new channel materials are required to compensate the loss in mobility encountered at advanced CMOS technology nodes due to higher channel doping and scaled gate dielectrics. Strained Si technology can provide a solution for aggressively scaled devices by enhancing electronic transport properties. Electrons and holes have increased mobility in tensile strained Si compared with bulk Si. Therefore using a strained Si layer for the MOSFET channel can enhance device performance. Global strain across an entire wafer can be introduced using a virtual substrate of SiGe. The 4.2% lattice mismatch between Si and Ge results in tensile strained Si when epitaxially grown on relaxed SiGe. In global strain generation the amount of strain introduced is only limited by the critical thickness of a strained layer. Holes also benefit from higher carrier mobility in compressively strained SiGe. Compressive strain and multi-heterojunction devices such as resonant tunneling diodes and dual-channel CMOS, strained Si HBTs, etc. can also be realized using virtual substrate technology.

Performance gains of conventional virtual substrate devices are presently compromised by self-heating due to the low thermal conductivity of SiGe; the presence of just 1% Ge decreases the thermal conductivity by a factor of 4 [1]. This leads to poor power dissipation in an era where power consumption is becoming increasingly critical in determining the viability of a product. For strained Si analogue applications self-heating is an even greater challenge since devices consume power throughout their operation. The self-heating effect in strained Si/SiGe technology is aggravated since virtual substrates are conventionally several microns thick in order to reduce surface roughness and defect density to sufficiently low levels and to achieve the maximum level of strain relaxation possible. However, recent advances in silicon epitaxy have now enabled the production of high quality thin SiGe layers. By reducing the thickness of the virtual substrates self-heating effects can be minimized. Thinner SiGe layers also improve the economical viability of virtual substrate technology, both through reduced material consumption and faster epitaxial growth time. Nevertheless, in order to realize the potential advantages of thin virtual substrates, Si channel strain and defectivity must provide sufficient resilience to high temperature MOSFET processing. In thick virtual substrates it is relatively easy to achieve a high degree of relaxation in the SiGe layer, thus channel strain and the critical thickness to relaxation is well understood. However there is little published work on the thermal stability of strained Si layers grown on thin virtual substrates. In particular the effect of strained Si channel thickness, a key parameter in MOSFET design, on thin virtual substrate material has not previously been considered. Raman spectroscopy is a fast and non-destructive method for quantifying strain in thin layers. Inelastic scattering of light due to lattice vibrations and electronic excitations is measured and a frequency shift in the reflected light compared with unstrained material relates to the residual or applied strain. In this paper Raman spectroscopy is used to investigate the thermal stability of strained Si layers grown on thin virtual substrates at thicknesses above and below the critical thickness.

EPITAXIAL GROWTH METHOD AND PROCESSING

The structures used in the study are shown in Fig. 1. All layers were grown in a 200 mm ASM Epsilon reactor at 600 °C. High levels of relaxation in thin SiGe layers were achieved by incorporating a thin C-containing layer into the constant composition $Si_{0.78}Ge_{0.22}$ layer [2]. C stimulates relaxation by decreasing the energy barrier for dislocation formation. Previous work has shown that 75% relaxation is achieved after growth of a second C-free $Si_{0.78}Ge_{0.22}$ layer. Further relaxation is achieved after annealing at 1000 °C for 30 s. A strain adjustment layer containing 18% Ge is subsequently grown. The lower Ge content is intended to match the lattice constant of the $Si_{0.78}Ge_{0.22}$ virtual substrate. Strained Si channel thicknesses of 12, 30 and 80 nm were grown. Since the critical thickness of Si on relaxed SiGe is approximately 15 nm [3], only the 12 nm layer was expected to be fully coherent with the underlying virtual substrate. Layer thicknesses were confirmed using spectroscopic ellipsometry. A thick virtual substrate with a 12 nm strained Si layer was also grown as a control using the same method.

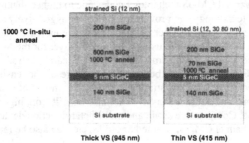

Fig. 1. Epitaxial layer structures.

Channel strain, defect density and surface roughness were assessed on as-grown samples. Channel strain was determined using a 514.5 nm laser fitted to a Jobin Yvon LabRAM HR system. Defect density was analysed using a modified Schimmel etch and surface quality was studied by AFM. Samples from each wafer were then subjected to an inert anneal performed at 800, 900, 1000 or 1050 °C for 30 s. Such anneals are typical of the highest temperature fabrication stage encountered during MOSFET processing, which is the source/drain diffusion activation. Following annealing material properties were measured again in order to evaluate thermal stability. Initial electrical characterisation was carried out on n-MOSFET devices fabricated on a thin virtual substrate having a 10 nm stained Si channel layer.

RESULTS

In virtual substrate devices the Raman peak due to the Si-Si vibrations in the strained Si layer is partially hidden by the large peak due to the Si-Si vibrations in the relaxed underlying SiGe layer. For relatively high Ge content virtual substrates the individual peaks are detectable but for lower virtual substrate Ge contents or thin strained Si layers the peak positions, which relate to the layer strain, become increasingly difficult to resolve using a 514.5 nm laser. In order to accurately determine the peak position of the weaker Raman response from the Si channel, a peak-fitting approach was used. The peak-fitting procedure was calibrated using a selective etch to remove the strained Si layer from the virtual substrate [4]. This enabled the Si channel strain to be evaluated as a function of annealing temperature. The Si peak position for each sample is shown in Fig. 2 below. The spectral resolution of the Raman microscope is +/- 0.2 cm^{-1}. The larger error bars shown on Fig. 2 arise from the curve fitting procedure. However the error shown (+/- 0.5 cm^{-1}) is the worse case scenario and the true error in the data is likely to be much smaller. Fig. 2 shows that for samples above the critical thickness (30 nm, 80 nm) the levels of strain in the Si layers are lower than for the 12 nm sample. This is evident by the peak position of the super-critical thickness layers being located at higher wavenumbers than that of the sub-critical thickness sample. Although up to 70% of the strain is maintained in the thick Si layers, additional relaxation occurs following annealing. A distinct increase in the peak position is observed for the 30 nm layer after annealing at 1050 °C and for the 80 nm layer after annealing at 900 °C. Supercritical strained Si layers on thin virtual substrates must therefore be considered unstable following MOSFET processing. The earlier onset of additional strain relaxation in the 80 nm sample compared with the 30 nm sample is due to the 80 nm layer exceeding the critical thickness by a larger amount.

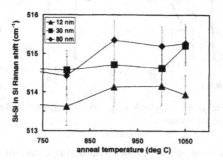

Fig. 2. Variation in Raman peak position with annealing temperature for sub- (12 nm) and super-critical (30 nm, 80 nm) strained Si layers on thin virtual substrates.

Fig. 3 compares the thermal stability of 12 nm Si layers grown on thin and thick virtual substrates. There is no consistent trend of strain with annealing temperature, indicating that both layers provide similar resilience against high thermal budget processing. However the downward shift of the peak position for the Si layer grown on the thick virtual substrate compared with the thin virtual substrate indicates that the thick virtual substrate is providing the Si with more strain. Since virtual substrate growth differed only in thickness these results suggest that relaxation and therefore the amount of strain induced in the Si is compromised in thin virtual substrates.

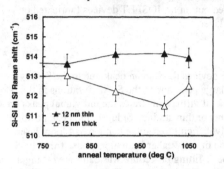

Fig. 3. Raman peak position for 12 nm strained Si layers on thin and thick virtual substrates.

The Raman signal collected from the Si channel compared with the signal collected from the virtual substrate can be used to give an indication of layer thickness because as the strained Si layer thickness is increased, a greater proportion of the Raman signal comes from the surface Si [Fig. 4(a)]. Since the thickness of the thin virtual substrates was the same in all samples, the relative intensities of the peak due to the Si channel (which varied in each sample) and the peak due to the SiGe virtual substrate correlates with the Si layer thickness [Fig. 4(b)]. Therefore potential changes in strained layer thickness due to Ge out-diffusion from the SiGe during processing can be studied.

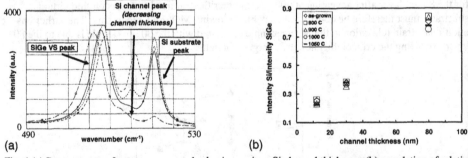

Fig. 4.(a) Raman spectra for as-grown samples having various Si channel thickness; (b) correlation of relative intensities (Si/SiGe) with channel thickness.

Fig. 5 shows the relative intensity of the strained Si layer and the relaxed SiGe layer, normalised to as-grown values (prior to annealing). A general decrease in the relative intensity with increasing annealing temperature is observed due to Ge diffusion. Using calibrated TCAD simulations the maximum Ge diffusion length (i.e. after the 1050 °C anneal) is predicted to be 2 nm. This change will have the greatest impact on the thinnest channels (12 nm). Although the largest decrease in relative intensity with temperature occurs in the 12 nm layer, the lack of clarity in the trends suggests that further calibration of the Raman signal intensity is required in order to detect such small variations in layer thickness. SIMS was also unable to detect any change in these layer thicknesses due to Ge diffusion until 1100 °C annealing was performed.

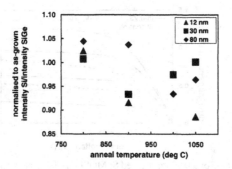

Fig. 5. Relative Raman peak intensities for strained Si and relaxed $Si_{0.78}Ge_{0.22}$ layers.

Defect etching was carried out on all samples and several trends were identified. The as-grown threading dislocation density was found to increase with increasing channel thickness. The dislocation density also increased with increasing annealing temperature, even for thin channels. Since negligible relaxation was identified by Raman in thin Si channels following any anneal, these results indicate that Schimmel etching may provide increased sensitivity to strain relaxation (which introduces defects) than use of the Raman peak position. This has also been observed for devices fabricated on thick virtual substrates [5].

The unetched material quality was further investigated by AFM. Surface defects approximately 40 nm in peak height were revealed on supercritical strained layers (Fig. 6). These features increased in density as the strained Si layer thickness increased. Consequently although significant macrostrain is maintained in supercritical layers, the work demonstrates that micro-Raman alone cannot determine the suitability of material for device performance.

Fig. 6. 10 μm x 10 μm 3D AFM image of supercritical strained Si layer on thin $Si_{0.78}Ge_{0.22}$ virtual substrate.

Surface defects were not evident on the sub-critical strained layers and n-MOSFET devices were fabricated on a thin $Si_{0.78}Ge_{0.22}$ virtual substrate with a 10 nm Si layer. Output characteristics for 0.13 μm gate length devices and Si control devices are shown in Fig. 7. The increase in drain current compared with the Si control is almost 20% at $V_d = V_g-V_t = 1.0$ V, verifying that significant performance gains are possible using thin virtual substrate material. Moreover, the constant drain current observed after saturation indicates that the impact of self-heating is negligible. This contrast previous reports of thick SiGe virtual substrate devices, in which performance enhancements are severely compromised by self-heating [6]. Therefore thin virtual substrates generated using C-incorporation provide an effective solution for increasing device performance while avoiding self-heating and lengthy epitaxial growth times.

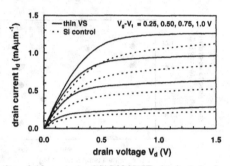

Fig. 7. Output characteristics for a 0.13 μm strained Si MOSFETs fabricated using a 10 nm Si layer on a thin $Si_{0.78}Ge_{0.22}$ virtual substrate.

CONCLUSIONS

Good stability during high temperature processing has been demonstrated for strained Si layers grown below the critical thickness on thin virtual substrates. The $Si_{0.78}Ge_{0.22}$ virtual substrates were grown using C to promote relaxation in SiGe. Si layers grown above the critical thickness exhibit reduced strain and further relaxation following high temperature annealing. The annealing temperature at which the onset of strain relaxation occurred decreased with increasing layer thickness. Further relaxation was detected in 30 nm and 80 nm layers following anneals of 1050 °C and 900 °C, respectively. Strain in Si layers grown on thin and thick virtual substrates having identical Ge composition and epilayer thickness has also been found to correlate with the degree of relaxation achieved in the SiGe virtual substrate; for identical growth conditions thick virtual substrates are more relaxed than thin virtual substrates. Although significant macrostrain is maintained in super-critical strained layers following annealing, prominent surface defects develop. The defects were not apparent on sub-critical thickness strained layers and electrical measurements from n-MOSFETs fabricated on these layers display significant performance gains compared with bulk Si controls and reduced self-heating effects compared with thick virtual substrate devices. Raman data have been used to study the Si strained layer thickness and Ge out-diffusion during high temperature processing. The work demonstrates that thin virtual substrates can provide suitable material properties and thermal stability to enable high performance devices.

ACKNOWLEDGMENTS: This work is supported through the EU Network of Excellence SiNANO and the EU Human Potential Programme. The Engineering and Physical Sciences Research Council (UK) and the National Council of Science and Technology (Mexico) are also acknowledged.

REFERENCES

1. D. G.Cahill et al, *Phys. Rev. B* **71**, 235202 (2005).
2. R. Delhougne et al, *Appl. Surf. Sci.* **224**, 91 (2004).
3. D. J. Paul, *Adv. Mater.* **11**, 191 (1999).
4. P. Dobrosz et al, *Surface Coatings Tech.* **200**, 1755 (2005).
5. S. H. Olsen et al, *J.Appl. Phys.* **97**, 114504 (2005).
6. K. Rim et al, *IEEE Trans. Electron Devices*, **47**, 1406 (2000).

Mater. Res. Soc. Symp. Proc. Vol. 913 © 2006 Materials Research Society 0913-D02-10

Impact of Heavy Boron Doping and Nickel Germanosilicide Contacts on Biaxial Compressive Strain in Pseudomorphic Silicon-Germanium Alloys on Silicon

Saurabh Chopra[1], Mehmet C Ozturk[1], Veena Misra[1], Kris McGuire[2], and Laurie McNeil[2]

[1]Electrical and Computer Engineering, North Carolina State University, Raleigh, NC, 27695

[2]Physics and Astronomy, UNC, Chapel Hill, NC, 27599

ABSTRACT

In recent years, the semiconductor industry has increasingly relied on strain as a performance enhancer for both n and p-MOSFETs. For p-MOSFETs, selectively grown $Si_{1-x}Ge_x$ alloys in recessed source/ drain regions are used to induce uniaxial compressive strain in the channel. In order to induce compressive strain effectively using this technology, a number of parameters including recess depth, $Si_{1-x}Ge_x$ thickness (junction thickness), sidewall thickness, dopant density, dislocation density, and contact materials have to be optimized. In this work, we have studied the effects of heavy boron doping and self-aligned germanosilicide formation on local strain. Raman spectroscopy has been used to study the impact of heavy boron doping on compressive stress in $Si_{1-x}Ge_x$ films. Strain energy calculations have been performed based on Vegard's law for ternary alloys and the effect of boron on strain in $Si_{1-x-y}Ge_xB_y$ alloys modeled quantitatively. It will be shown that, owing to the smaller size of a boron atom, one substitutional boron atom compensates the strain due to 6.9 germanium atoms in the $Si_{1-x-y}Ge_xB_y$ film grown pseudomorphically on silicon. The critical thickness of $Si_{1-x-y}Ge_xB_y$ has been calculated for the first time based on kinetically limited critical thickness calculations for metastable $Si_{1-x}Ge_x$ films. It will be shown that the critical thickness of the alloy increases as the boron content in the alloy is increased, making boron concentration an additional parameter for optimizing strain in the MOSFET. Based on these conclusions, boron concentration can be used to preserve the strain for thicker $Si_{1-x-y}Ge_xB_y$ films (compared to $Si_{1-x}Ge_x$ films) while keeping the dislocation density low. Furthermore, we show that NiSiGe contacts can have a profound impact on the SiGe strain. Our results indicate that NiSiGe introduces additional stress in the underlying $Si_{1-x-y}Ge_xB_y$, which further affects the strain induced in the channel.

INTRODUCTION

In applications where $Si_{1-x}Ge_x$ is heavily doped with boron, smaller boron atoms can partially compensate the strain in the alloy. This effect is similar to that of carbon in $Si_{1-x-y}Ge_xC_y$ alloys. In this work, we have investigated the impact of boron on biaxial compressive strain in $Si_{1-x}Ge_x$ alloys pseudomorphically grown on Si. The stress in the epitaxial layers was characterized by Raman spectroscopy and theoretical calculations. Using boron strain compensation, the Ge content of the alloy can be substantially increased without increasing the stored strain energy. This phenomenon can be used to increase the critical thickness for a given Ge concentration, which can have many advantages. Critical thickness calculations for metastable $Si_{1-x-y}Ge_xB_y$ alloys based on Houghton's model [1] are reported for the first time as a function of boron and germanium concentrations. The calculations are further supported by strain energy density measurements obtained by Raman spectroscopy. In addition to these parameters, the contact metal may also impact the channel stress. In this work, we have studied the impact of nickel germanosilicide (NiSiGe) on the biaxial compressive stress in SiGe used in the junction regions and hence, the MOSFET channel. The results indicate that it is very important to optimize all of these parameters since they have important implications in source/drain engineering of nanoscale MOSFETs with recessed source/drain $Si_{1-x}Ge_x$ junctions.

EXPERIMENTAL DETAILS

In this study, 150 mm, (5 Ω.cm) n-type Si wafers of (100) orientation were used. Samples for Raman Spectroscopy were prepared by selective epitaxy of in-situ boron doped $Si_{1-x}Ge_x$ alloys in windows defined in a thermally grown oxide layer (~100 nm) by photolithography and wet chemical etching. Epitaxy was performed in a single wafer, cold-walled ultra high vacuum rapid thermal chemical vapor deposition (UHV-RTCVD) system [2]. The $Si_{1-x}Ge_x$ layers were grown with different germanium concentrations and varying in-situ doping levels of boron at 500 and 550 °C. The boron concentration in the epitaxial layers was within 0.5 to 2 %; hence, they will be referred to as ternary $Si_{1-x-y}Ge_xB_y$ alloys. The layer thickness was determined using atomic force microscopy (AFM) by measuring the step height between the selectively grown layer and the surrounding oxide. The sheet resistance of the layers was measured using a four point probe apparatus (Magnetron M-700). The germanium and boron concentration in the films was determined by secondary ion mass spectroscopy (SIMS). Raman analysis was carried out using a micro-Raman manufactured by Dilor systems (λ = 488 nm). The amount of strain in $Si_{1-x-y}Ge_xB_y$ was calculated by comparing the peak shifts in the Raman spectra of these films [3]. In order to study the effect of germano-silicides on the $Si_{1-x}Ge_x$ regions, samples were prepared by sputtering Ni (~10 to 30nm) on $Si_{1-x}Ge_x$ layers and then annealing at 450° C for 30sec using a rapid thermal anneal system (Heat Pulse 210A). The excess Ni was removed using a 1:4 H_2O_2: H_2SO_4 solution which etches Ni selectively with respect to the germanosilicide.

DISCUSSION

Germanium concentration in the epitaxial $Si_{1-x}Ge_x$ layers was measured by both Raman spectroscopy and SIMS and the results are summarized in table I. Germanium concentration in the SIMS profiles was established by analysis under identical conditions of a set of $Si_{1-x}Ge_x$ standards. The Ge concentration in these standards was established by Rutherford Backscattering Spectrometry (RBS). Raman analysis was performed on fully relaxed $Si_{1-x}Ge_x$ layers grown above the critical thickness [1]. As shown, the two techniques are in close agreement. Table II provides a summary of the measured Si-Ge peak positions for relaxed and strained $Si_{1-x}Ge_x$ layers with different germanium and boron concentrations. The relative strain energy in the epitaxial layers was determined by measuring the shifts of the Si-Ge peaks from their positions in relaxed $Si_{1-x}Ge_x$ [3]. The thickness of the strained $Si_{1-x}Ge_x$ layers was kept well below the measured critical thickness for all the germanium concentrations studied (thickness ~ 25 nm). In calculating the strain, we assumed that the boron atoms occupied substitutional sites. It is understood that this can be a source of error; however, we believe that the observed trend would still be valid.

Table I. Germanium concentration in $Si_{1-x}Ge_x$ layers determined by Raman Spectroscopy and SIMS

Si-Si Peak (cm^{-1})	% Ge (Raman)	% Ge (SIMS)
487.5	47.8	45.2
502.6	28.4	27.8
509.3	17.6	17.5

Table II. $Si_{1-x}Ge_x$ phonon peaks obtained from Raman Spectroscopy on $Si_{1-x-y}Ge_xB_y$ samples

Boron concentration (cm^{-3})	Ge = 50%		Ge = 28%		Ge = 17%	
	Relaxed	Strained	Relaxed	Strained	Relaxed	Strained
1 X 10^{21}	405.3	407.6	406.7	407.2	405.1	405.1
4.8 X 10^{20}	405.3	408.4				
5.8 X 10^{19}	405.3	409.3				
undoped	405.3	409.5	406.7	408.2	405.1	405.6

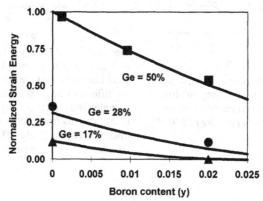

Figure 1. Normalized Strain energy as a function of boron content. The solid lines represent the theoretical calculations and the symbols represent the experimental data as a function of boron content for three different germanium concentrations.

In order to model the effect of boron on strain in $Si_{1-x-y}Ge_xB_y$, we need to calculate the lattice mismatch between the alloy and the substrate. The alloy lattice constant was determined using Vegard's law for ternary alloys [4]. Since boron does not exist in a cubic structure like silicon or germanium, the boron lattice constant was extrapolated from the ratio of the covalent radii of Si, Ge and B [5]. Assuming pseudomorphic $Si_{1-x-y}Ge_xB_y$ growth on Si, the strain in the alloy was estimated using the misfit parameter [6]. Assuming that there are no dislocations at the $Si_{1-x-y}Ge_xB_y$/Si interface, the stored strain energy in the film, can be calculated as outlined in a previous publication by this laboratory [5]. Figure 1 shows the strain energy as a function boron concentration for three different Ge concentrations. Solid lines represent the results of the strain energy calculations and data points represent the values obtained from Raman measurements. It can be seen that the measurement results are in close agreement with the theoretical predictions. For fully compensated films (i.e. zero strain) the Ge:B ratio is approximately 6.9, which compares well with the ratio of 6.45 calculated by Maszara et al.[7]. According to Figure 1, boron strain compensation can be used to reduce the strain energy for a given Ge concentration. Alternatively, the same strain energy can be obtained at a higher Ge concentration by increasing the boron concentration. Consequently, thicker layers with higher Ge concentrations can be grown effectively increasing the critical thickness.

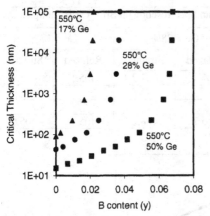

Figure 2. Kinetically limited critical thickness of a $Si_{1-x-y}Ge_xB_y$ alloy grown using an ultra high vacuum rapid thermal chemical vapor deposition system at a temperature of 550 °C.

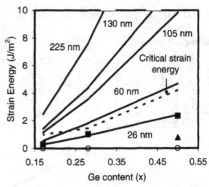

Figure 3. Strain energy density stored in a $Si_{1-x}Ge_x$ alloy as a function of Ge content (x). The solid lines represent the theoretical calculations for different film thicknesses. The squares, triangles and circles correspond to measurements from ~ 26nm, 60nm, 105nm thick films respectively. The broken line represents the critical strain energy density.

Critical thickness characterization:

In fabrication of p-channel MOSFETs with recessed $Si_{1-x}Ge_x$ junctions, an important parameter that needs to be optimized is the thickness of the selectively deposited $Si_{1-x}Ge_x$ in the junction areas. If the $Si_{1-x}Ge_x$ is not thick enough, the junctions fail to induce the required amount of stress in the channel region. On the other hand, the junctions can not be made very thick because the films begin to relax via formation of misfit dislocations beyond a certain thickness. The thickness at which complete relaxation occurs is called the critical thickness. Presence of misfit dislocations in the junction areas lead to increased leakage as well as reduction in the channel stress. Therefore, it is very important to accurately determine the critical thickness for a system and optimize the process to reduce the number of dislocations and maximize the channel stress. In the present case, critical thickness for $Si_{1-x-y}Ge_xB_y$ alloys on silicon was calculated using the model for kinetically limited critical thickness proposed by Houghton [1]. This is plotted as a function of boron concentration for three different germanium concentrations in figure 2. According to this model, strain relaxation is a strong function of film thickness and various thermal cycles in the process. Raman spectroscopy was again used to determine the strain energy in $Si_{1-x}Ge_x$ films of varying thicknesses and this experimental data was compared with the calculated meta-stable critical thickness in figure 2. Figure 3 shows the measured strain energy and corresponding theoretical results as a function of the germanium content, for zero boron content (undoped films). The lines represent the calculated results, and the symbols represent the measured results. The broken line represents the calculated critical strain energy density for undoped $Si_{1-x}Ge_x$ as a function of germanium thickness. Based on the calculated strain energy, samples which fall below the critical value should be strained while the ones which are above should undergo relaxation via misfit dislocation formation at the interface. From figure 3, it is observed that the experimentally determined strain energy of samples with t ~ 26nm (squares) coincide with the theoretically calculated values. For the sample with t ~ 60nm, the theoretical calculations predict the onset of relaxation in the film, since the calculated strain energy is above the critical value. This is confirmed by experimental results which show that the measured strain energy (triangles) is less than the calculated value due to strain relaxation in the film. The same trend is observed for the rest of the samples since all these samples have calculated strain energies greater than the critical energy.

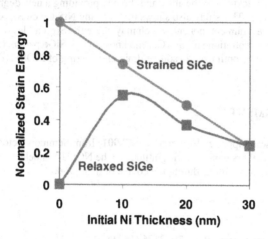

Figure 4. The normalized strain energy in $Si_{1-x}Ge_x$ as a function of Ni thickness after NiSiGe formation on relaxed and fully strained $Si_{1-x}Ge_x$.

Effect of Nickel Germanosilicide:

As mentioned above, the formation of NiSiGe on top of the $Si_{1-x}Ge_x$ layer can impact the strain in the layers. In order to study this effect, NiSiGe was formed on top of relaxed as well as strained $Si_{1-x}Ge_x$ layers. Raman spectroscopy was used to monitor the stress in $Si_{1-x}Ge_x$ under the NiSiGe. Figure 4 shows the normalized compressive strain energy as a function of the deposited nickel thickness (hence the NiSiGe thickness) for two different cases. In the first case (squares), the starting $Si_{1-x}Ge_x$ is relaxed as shown in Figure 4 for zero Ni thickness. After formation of the NiSiGe layer with 10 nm of Ni, the $Si_{1-x}Ge_x$ layer is compressively strained. This stress is entirely caused by the top NiSiGe layer since the misfit dislocations already exist at the $Si/Si_{1-x}Ge_x$ interface. As the NiSiGe thickness is increased, the increasing stress eventually causes relaxation to occur thereby decreasing the magnitude of stress measured in the $Si_{1-x}Ge_x$ layer. In the second case (circles), we begin with fully strained layers but close to the critical thickness for relaxation. Therefore, with the addition of the NiSiGe layer, the strain energy in $Si_{1-x}Ge_x$ can easily exceed the critical energy inducing relaxation and effectively decreasing the measured strain energy with increasing NiSiGe thickness. This view is supported by Zhao et al. in a recent report [8]. The results show that optimizing the NiSiGe parameters is essential in source/drain engineering of MOSFETs with $Si_{1-x}Ge_x$ source/drain junctions.

CONCLUSIONS

This study shows that boron incorporation in $Si_{1-x}Ge_x$ alloys grown pseudomorphically on silicon can reduce the biaxial compressive stress in the alloy. One boron atom can compensate 6.9 Ge atoms. It was also shown that increasing boron concentration can be used to increase the critical thickness of the alloy and thereby providing a new degree of freedom in band-gap engineering. This study also shows that forming NiSiGe on $Si_{1-x}Ge_x$ can introduce another compressive strain component, which may cause the $Si_{1-x}Ge_x$ layer to relax. In summary, in addition to Ge concentration and $Si_{1-x}Ge_x$ thickness, the NiSiGe parameters and boron concentration should be optimized in source/drain engineering of MOSFETs with recessed $Si_{1-x}Ge_x$ junctions.

ACKNOWLEDGMENTS

The work was supported by a grant (**1137.001**) from Semiconductor Research Corporation. The authors express their gratitude to the NCSU Nanoelectronics Facility personnel for their contributions during the course of their work.

REFERENCES

1. D. Houghton, *J. Appl. Phys.* **70**, 2136 (1991).
2. J. Liu, M.C. Ozturk, *IEEE Trans. Elec. Dev.* **52**, 1535 (2005).
3. W. Byra, *Solid State Comm.* **9**, 2271 (1971).
4. S. Iyer, K. Eberl, M. Goorsky, F. LeGoues, J. Tsang, F. Cardone, *Appl. Phys. Lett.* **60**, 356 (1992).
5. S. Chopra, M. C. Ozturk, V. Misra, K. McGuire, L. McNeil, to be published in *Appl. Phys. Lett.*
6. *Properties of Silicon Germanium and SiGe:Carbon*, ed. E. Kasper, K. Lyutovich (INSPEC-IEE, 2000).
7. W. Maszara, T. Thompson, *J. Appl. Phys.* **72**, 4477 (1992).
8. Q.T Zhao, D. Buca, St. Lenk, R. Loo, M. Caymax, S. Mantl, *Microelectronic Engineering*, **76**, 285 (2004)

Poster Session

Mater. Res. Soc. Symp. Proc. Vol. 913 © 2006 Materials Research Society

Evidence of Reduced Self Heating with Partially Depleted SOI MOSFET Scaling

Georges Guegan[1], Romain Gwoziecki[1], Olivier Gonnard[2], Gilles Gouget[2], Christine Raynaud[1], Mikael Casse[1], and Simon Deleonibus[1]

[1]CEA-DRT-LETI, GRENOBLE Cedex 9, 38054, France
[2]STMicroelectronics, CROLLES, 38920, France

ABSTRACT

SOI technology offers advantages over bulk silicon and become a good candidate for analog/RF applications. However, the presence of a buried oxide causes self-heating which can degrade the device performance. The effects of self-heating have been examined on two successive generations with a wide range of device dimensions. This work shows that self-heating effects become less and less severe with both MOSFET and power supply voltage scaling. Moreover, we have observed a weak increase of 100 nm pMOSFET output characteristics due to the self-heating.

INTRODUCTION

The advantages of SOI technology, high speed, low power dissipation, co integration of digital and analog/RF circuits [1], arise from the presence of a buried oxide in silicon. However, this oxide layer increases thermal insulation of the channel region due to the smaller thermal conductivity of silicon dioxide compared to that of bulk silicon. On one hand, the silicon device scaling not only shortens design rules but also silicon film thickness is reduced. Therefore, the self heating-effect (SHE), which is an important issue for SOI CMOS, may become more pronounced because of increased electric field density and reduced silicon volume available for heat removal [2]. On the other hand, the channel temperature may be reduced as a result of both buried oxide thickness and power supply voltage scaling.

Thus this paper examines the impact of both devices and supply voltage scaling on SHE of both 130 nm and 90 nm SOI technologies. Furthermore, thermal resistances of these two CMOS SOI technology generations and self heating DC parameters have been extracted and compared. The relative contributions of key parameters on the body contacted and floating body current degradation induced by self heating have been quantified with BSIMSOI model for both PD SOI pMOSFET and nMOSFET devices.

DEVICE STRUCTURE AND MEASUREMENT TECHNIQUE

SOI devices of 130 nm and 90 nm nodes are partially depleted SOI MOSFET with either body-contacted (BC-SOI) or floating-body (FB-SOI) MOSFET manufactured respectively on 200 mm and 300 mm Unibond® wafers. Main geometrical parameters and power supply voltage of these two successive SOI CMOS technologies are compared on Table 1.

Table I. Supply voltage and device dimension.

Nodes	130 nm	90 nm
Supply voltage	1.2 volt	1 volt
Drawn gate length	130 nm	100 nm
Initial silicon film thickness	160 nm	70 nm
Buried oxide thickness	400 nm	145 nm

The channel temperature measurements and the impact of thermal heating on partially depleted SOI device performances were investigated with gate resistance thermometry technique [3]. Test structures have been defined for the temperature measurement with gate designed for two point resistance measurement. The substrate temperature was monitoring with variable temperature chuck (25°C – 125°C). Figure 1 represents an example of polysilicon gate temperature calibration when the device is turn-off (Vd=0 V). Then the device is turn on and the temperature increase is estimated from gate resistance and calibration measurements. Figure 2 gives the channel temperature enhancement versus the power extracted on a 100 nm gate length FB-SOI nMOSFET. The thermal resistance is given by the slope of the channel temperature versus device power.

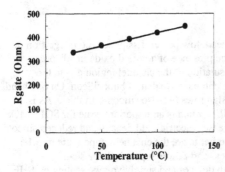

Figure 1. Gate temperature calibration of a FB SOI nMOSFET (W/L=1 µm/100 nm).

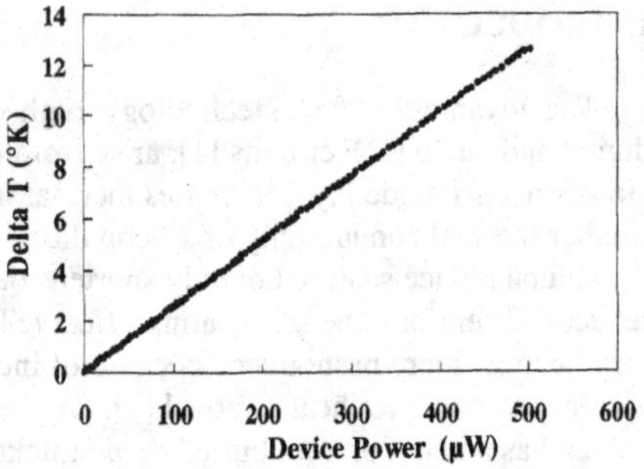

Figure 2. Channel temperature increase versus device power of a FB SOI nMOSFET (W/L=1 µm/100 nm).

THERMAL RESISTANCE COMPARISON

As seen in figure 3, the maximum temperature rise at the nominal supply voltage increases with decreasing gate length and increasing channel width. On the other hand, the thermal resistances increase when gate length and channel width decrease, due to the substrate surface reduction. Figure 4 compares n and p channel temperature enhancement of similar device geometry W/L=320 nm/100 nm defined with either BC or FB device.

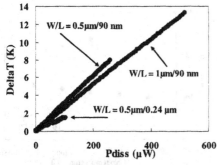 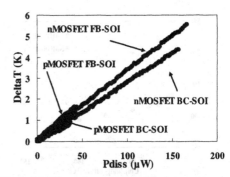

Figure 3. Channel temperature increase versus device power for several FB SOI nMOSFET geometries.

Figure 4. Channel temperature increase versus device power for FB and BC SOI MOSFET (W/L=320 nm/100 nm).

We can notice that the thermal resistance of NMOS and PMOS are logically equals while device temperature increases with nMOSFET due to higher driving capability. It is apparent that the FB-SOI thermal resistances are larger than that of the BC-SOI devices for the nominal 100 nm gate length. This effect may be attributed to the design of FB device which leads to silicon volume decrease.

ANALYSIS OF THE SELF HEATING EFFECT ON DRAIN CURRENT LEVEL

The impact of thermal heating on the degradation of SOI device drain current level has been investigated from Id-Vd characteristics monitored at variable chuck temperatures. The dependences of nMOSFET and pMOSFET saturated drain current on the chuck temperature are respectively compared on figures 5 and 6 for the 90 nm node devices. The drain current dependence of FB nMOSFET with chuck temperature is higher than for BC devices (see figure 5). In contrast to the standard behaviour, we observe an increase of BC pMOSFET saturated drain current with temperature for either 110 nm (figure 6) or 100 nm gate length (figure 7). As a general rule, the FB-SOI devices are more sensitive to the SHE than the BC-SOI devices whatever channel widths are, as shown in Figure 8. This behaviour is due to both higher channel temperature of FB devices and floating body effects which tend to decrease with temperature. These experimental results also demonstrate the weak sensitivity of p channel MOSFET to SHE due to the lower current which results in a smaller power dissipation and a lower channel temperature enhancement.

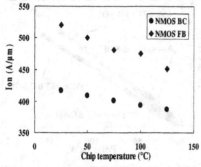

Figure 5. Temperature dependency of Ion for 110 nm FB and BC-SOI nMOSFET devices.

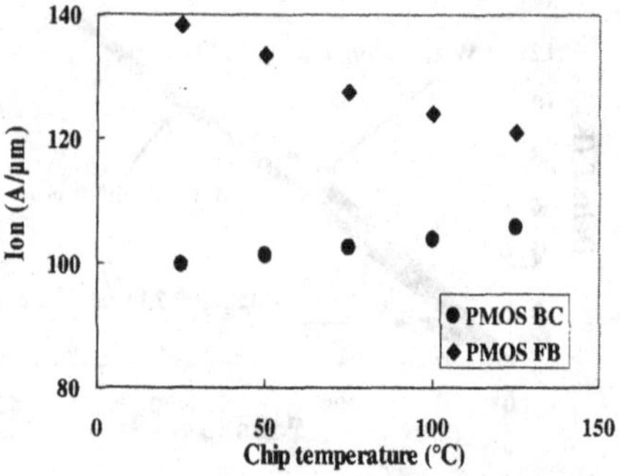

Figure 6. Temperature dependency of Ion for 110 nm FB and BC-SOI pMOSFET devices.

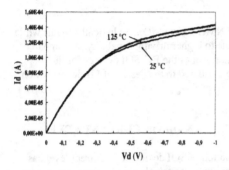

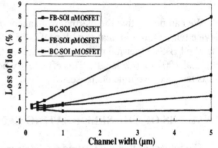

Figure 7. Comparison of the output characteristics of 100 nm BC pMOSFET at 25°C and 125°C.

Figure 8. Variation of Ion gain or loss with channel width of either FB or BC SOI MOSFET at 1 volt (L=100 nm).

In order to clarify these experimental results, a temperature-dependence compact model was used to investigate and to quantify the effect of key parameters on drain current level. The carrier mobility, the threshold voltage and the carrier saturation velocity are the main DC MOSFETs parameters which are temperature dependent [4]. In addition, the kink effect which is temperature dependent due to impact ionization and recombination current I_R, must be considered for SOI floating-body devices. The recombination current which is more temperature-sensitive than impact-ionization mechanism tends to reduce the kink effect with temperature [5]. The popular SPICE BSIM3SOI2.2.3 model [6] has been used to investigate current modulation due to the SHE. Usually, the drain current of SOI device is reduced as a result of lower mobility due to higher temperature. The relative contribution of both mobility and threshold voltage are gate length dependent. Simulations with BSIM SOI model allow to explain the paradoxical behaviour which has been observed in the case of p channel BC-SOI device. The lower saturated threshold voltage at high temperature may counteract the effective mobility degradation. The total current degradation due to self heating and the relative contributions of key parameters extracted with BSIMSOI model of either nMOSFET or pMOSFET are shown respectively on figures 9 and 10. The relative contributions of each parameter are higher with FB devices due to the higher local temperature. The effect of carrier saturation velocity is negligible.

The Ion degradation of nMOSFET due to SHE of both 130 nm and 90 nm nodes are compared on figure 11. The less significant degradation of 90 nm node devices is due to both lower device temperature rise and lower thermal resistance (figure 12).

RTH, which is the normalized thermal resistance, has been calculated for each kind of device. The dependence of RTH with geometry is modelled with this expression in BSIMSOI model [6]: RTH = (RTH0 + LRTH0/Leff) / (Weff +WTH0) where Weff and Leff are the effective channel width and channel length. RTH0, LRTH0 and WTH0 are BSIM binning parameters. Figures 13 and 14 illustrate the dependence of the normalized thermal resistance on channel width and gate length for the 90 nm node devices. The comparisons between these measurements and the model of RTH after extraction of binning parameters are also shown.

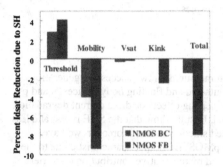

Figure 9. Relative contribution of 100 nm nMOSFET key parameters on the Ion degradation induced by self heating.

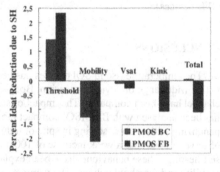

Figure 10. Relative contribution of 100 nm pMOSFET key parameters on the Ion degradation induced by self heating.

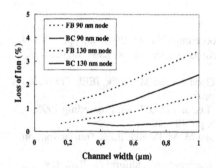

Figure 11. Variation of Ion loss with channel width of the nominal gate length for FB and BC SOI nMOSFET.

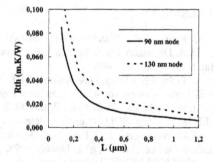

Figure 12. Comparison between 130 nm and 90 nm node normalized thermal resistance (floating body devices; W=5 μm).

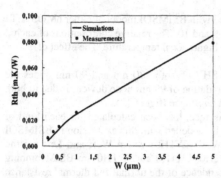

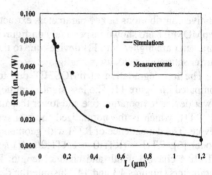

Figure 13. Extracted normalized thermal resistance dependence with channel width and comparison with RTH compact model (FB devices).

Figure 14. Extracted normalized thermal resistance dependence with gate length and comparison with RTH compact model (FB devices).

CONCLUSIONS

The temperature rise in SOI devices has been measured on two successive generations with a wide range of device dimension. Body-contacted and floating body devices, n and p channel have been compared. The impact of self heating effects on drain current degradation has been analysed with BSIMSOI compact model. Results show that the SHE is less and less penalizing with process scaling in spite of reduced silicon volume. Moreover, we have observed in our case a weak increase of 100 nm pMOSFET output characteristic due to the self heating. These behaviours have been explained by the relative contribution of carrier mobility and threshold voltage. The parameters of the thermal resistances have been extracted and the dependence of the thermal resistance with device dimensions has been investigated with experiments and BSIMSOI RTH model.

REFERENCES

1. C. Raynaud et al, Electrochemical Society Proceedings Volume 2005-03, pp. 331-344.
2. D. A. Dallmann and K. Shenai, IEEE Transactions on Electron Devices, vol. 42, n° 3, March 1995, pp. 489 – 496.
3. L. T. Su, J.E. Chung, D.A. Antoniadis, K. E. Goodson and M. I. Flik, IEEE Transactions on Electron Devices, vol. 41, n° 1, January 1994, pp. 69 – 75.
4. B. M. Tenbroek, W. Redman-White, M. S. L. Lee and M. J. Uren, Proceeding 1995, IEEE International SOI Conference, October 1995, pp. 48 – 49.
5. G. O. Workmann, J. G. Fossum, S. Krishnan and M. M. Pelella, IEEE Transactions on Electron Devices, vol. 45, n° 1, January 1998, pp. 125 – 133.
6. BSIM3SOI MOSFET Model Users' Manual, University of California at Berkeley, February 2002.

Mater. Res. Soc. Symp. Proc. Vol. 913 © 2006 Materials Research Society 0913-D03-03

Quantum Well Nanopillar Transistors

Shu-Fen Hu[1], and Chin-Lung Sung[2]
[1]RDT, National Nano Device Laboratories, 26, Prosperity Road I, Science-based Industrial Park, Hsinchu, Taiwan, 30078, Taiwan
[2]National Nano Device Laboratories, 26, Prosperity Road I, Science-based Industrial Park, Hsinchu, Taiwan, 30078, Taiwan

ABSTRACT

We have fabricated vertical quantum well nanopillar transistors that consist of a vertical stack of coupled asymmetric quantum wells in a poly-silicon/ silicon nitride multilayer nanopillars configuration with each well having a unique size. The devices consist of resonant tunneling in the poly-silicon/ silicon nitride stacked pillar material system surrounded by a Schottky gate. The gate electrode surrounds half side of a silicon pillar island, and the channel region exists at all the pillar silicon island. Current-voltage measurements at room temperature show prominent quantum effects due to electron resonance tunneling with side-gate. Accordingly, the vertical transistor offers high-shrinkage feature. By using the occupied area of the ULSI can be shrunk to 10% of that using conventional planar transistor. The small-occupied area leads to the small capacitance and the small load resistance, resulting in high speed and low power operation.

INTRODUCTION

Confinement of electrons in a zero-dimensional semiconductor quantum dot has opened the way to a whole new class of experiments. As smaller and smaller quantum dots have been created, both the energy scale associated with the Coulomb blockade and the spacing of the single-particle eigenstates has increased. Recently, the developments in vertical transistor structures have produced devices with the potential for scaling down to the nanometer regime.[1, 2] Coulomb blockade effects are likely to become important as devices are scaled down and these have also generated a great deal of interest recently because of potential applications in new devices. Investigations of electron transport in nano-pillars in GaAs/GaAlAs heterostructures with well-defined barriers have shown Coulomb blockade oscillations as individual electrons are added to a quantum dot.[3] Similar structures in silicon are attractive because they are compact, have high, well-defined barrier heights, and are compatible with silicon technology. The vertical structure allows high packing density and the high controllability of layer thicknesses allows formation of an exact number of layers of conducting and insulating materials in which islands of well-defined dimensions and tunnel barriers with high, well-defined barrier heights can be formed. They also have advantages over lateral structures in disordered materials which show Coulomb blockade but do not always give a well-defined and reproducible number of islands and also over patterned nano-structures produced by lithography which present considerable fabrication challenges to obtain reproducible geometries and barrier heights. Fukuda et al.[4] demonstrated that ultra thin Si_3N_4 barriers can be formed in silicon pillars.

Fukuda et al.[5] demonstrated that ultra thin Si_3N_4 barriers can be formed in silicon pillars. D. M. Pooley et al.[6, 7, 8] showed the current–voltage characteristics of nano-pillars of polycrystalline silicon with two 2–3 nm thick silicon nitride tunnel barriers. Pillars with diameters between 45 and 100 nm showed a Coulomb blockade region and Coulomb staircase at 4.2 K. Moreover, Electron transport in silicon nanopillars with zero, one, or two silicon nitride barrier layers of 2 nm thicknesses showed Coulomb blockade and wide zero current regions are observed for some devices with two silicon nitride tunnel barriers. P. A. Lewis et al.[9] demonstrated High-density silicon nanopillar cathodes which were fabricated using a self-assembling colloidal gold etch mask. They have characterized electron emission from individual silicon nanopillar cathodes using a STM tip as an anode to obtain $I-V$ characteristics over a range of decreasing anode–cathode separations.

In this study, we have fabricated vertically stacked pillars using similar silicon and silicon nitride multilayer structures, to form quantum well between ultra small tunneling junctions and the source-drain region. The poly silicon/Si_3N_4/poly silicon/Si_3N_4/poly silicon/Si_3N_4/poly silicon/Si_3N_4/poly silicon material system is designed with two, three and four tunnel barriers, each 3 nm thick, to provide confinement of electrons on one, two and three charge islands respectively. The structures are designed to observe single electron effect at room temperature.

EXPERIMENT

The wafer layer structure is produced by a combination of low pressure chemical vapor deposition (LPCVD) of poly-crystalline silicon (poly silicon) and thermal nitridation by heating the wafer to 620°C in ammonia (NH₃) in a LPCVD system. The two layers of *in situ* phosphorus poly silicon are deposited at two ends as source and drain regions. The poly silicon layers in between silicon nitride layers are undoped. Figure 1 is illustrated a cross section of multilayer with four silicon nitride barriers schematically and a high-resolution electron transmitted micrograph (HRTEM). The pillars are defined using electron beam lithography Leica Weprint 200 system in NEB22A resist. The lithography system has an accelerating voltage of 40 kV, a beam current density of 4 A/cm², and a beam diameter of about 20 nm. Etching is performed in LAM TCP 9400SE, using 5 ~ 20 mTore mixing in an atmosphere of Cl₂, O₂, HBr and SF₆ at 65°C. A 9-10 nm-thick gate oxide is grown at 925°C in oxygen, further reducing the pillar dimension. Figure 2 shows the schematically "one quantum well" and the SEM image of top view device structure which is one un-doped poly silicon thin layer between two Si₃N₄ thin barrier layers. The thickness of the un-doped poly silicon thin layer and the two Si₃N₄ thin barrier layers is 3 nm. We fabricate one, two, and three quantum well devices respectively. The materials of source and drain are *in situ* phosphorus poly silicon. Al metal is deposited and patterned to be one-half surrounding the pillar side-gate between a 9~10 nm thickness silicon dioxide and multi-layer structure pillar, to provide electrical contacts to the device. An HP4156B semiconductor parameter analyzer measured the characteristics of the device at 300 K in air.

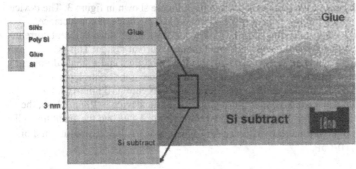

Figure 1. Schematic and high-resolution electron transmitted micrograph of the multiple poly silicon/nitride layers.

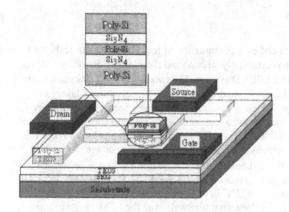

Figure 2. A schematic of the device structure is shown. Pillars with diameters from 10 nm were fabricated.

DISCUSSION

The current-voltage characteristic and differential conductance at room temperature for a 10 nm-diam pillar with two nitride barriers separated by 3 nm are shown in figure 3. The device exhibited a clear zero current region and staircase. There is clear evidence of conductance oscillations around the voltage extending to ± 0.7 V. The separaction between adjacent current steps is approximately 100 mV the same as the size of the zero current region. This is characteristic of Coulomb blockade in a system with two tunnel barriers and a single dominant island. [7, 10] We can estimate approximately the theoretical value of the control gate capacitance C_g. The thickness of the quantum well defined the structure is 3 nm. And, the diameter of the quantum well covered under the control gate of the device is 10 nm. Also, the thickness D of gate oxide, which is the distance between the control gate and the quantum well is 10 nm. Then the value of C_g estimated by $Cg = \varepsilon \varepsilon_0 A/D$, where relative dielectric constant ε of silicon dioxide is 3.9, is 0.1 aF.

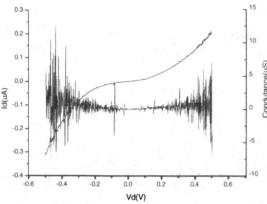

Figure 3. I-V and differential conductance characteristic of a two-barrier device at room temperature.

The current–voltage characteristics of most pillars with two or more Si_3N_4 barriers show a Coulomb gap and staircase and are not symmetrical at room temperature. Periodic peaks are seen directly in the current-voltage (I-V) characteristics in all the devices no matter one or two or three quantum well devices. The peaks exhibited a clear periodical oscillation. Furthermore, the oscillated frequency changed while the source-drain bias changed in the entire multi-barrier region. Increasing the applied voltage bias across the device progressively lowers the energy of all the states in the well relative to the energies the electrons in the source. Lowering the energy for all the states in the well and bringing them into resonance with the mobile electrons in the occupied conduction band in the source, so that an electron current can be transmitted through the device.

Quantum effects strongly influence the flow of electrons through these nanoelectronic devices. For the distance between the barriers is small (~3 nm), the widely space in energy level discrete and the 3 nm thin barriers cause electrons tunneling. A sequence of steps in current associated with each of the two energy scales is observed as the bias voltage is varied. The current jumps to a finite value when electrons can first travel through the island one at a time, and further large jumps herald the ability of electrons to go through two, then three at a time.

For one quantum well device (figure 4), consider the application of a bias voltage across the devices. At low voltage biasing, the device current is due to tunneling of electrons from filled states of a miniband in the top section to empty states in the corresponding miniband on the other side of the barrier. Increasing the bias voltage causes the current to rise steadily until the Fermi level on the top superlattice section side of the barrier falls within the lower minigap. When the second miniband on the topside becomes accessible to the tunneling electrons the current rises again.

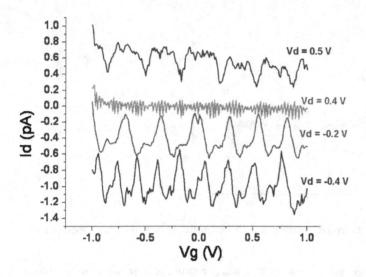

Figure 4. Current-voltage characteristics of a three-barrier device.

For two quantum well device, two dots with a strong inter dot-coupling form an artificial molecule with electronic states delocalized between the dots. The electron transport through the system is only possible when the levels diff from both wells are on resonance, and lie between the chemical potentials of the leads (in figure 5). The coherence of such extended states has been confirmed in recent experiments. [11, 12] In figure 4, clear periodic current oscillation can be seen, which persist to 300 K. The steps of voltage decrease when increase source-drain voltage. The decreased separations suggest effects of 3-dimention confinements in the triple barrier region, but the data is too complex to be explained by 3-dimention confinements in the quantum dot only. For three-quantum well device, the periodic oscillation frequencies do not change much while applied source-drain voltages as compare to one and two quantum well devices. The shift I-V curves can be observed when the source-drain bias increase. The cause of these current oscillations remain uncertain at present but it is possible they are the results of resonant tunneling and trap-assisted tunneling through trap states in the nitride or at the silicon-silicon nitride interface.[8] The overall current oscillation phenomena through the nanopillars are quite clear and to be able to observe stable periodic current oscillation in devices with four or more barriers.

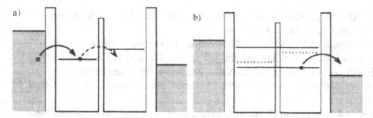

Figure 5. An electron can enter the double quantum dot from the left via a sequential tunneling processes indicated with the solid arrow. However, the electron is stuck in the left dot if the states in the two dots are not on resonance. b) For strong inter dot coupling, the states in the two dots (dotted) form hybridized molecular-like states (solid) that extend through the central barrier. These extended states enable electron transport through the system.[11]

CONCLUSIONS

Vertical resonant tunneling transistors with a Schottky gate have been fabricated and characterised with respect to their performance at room temperature. The devices show a very good peak current control by means of an applied gate voltage. The part-surrounding gate has a large field effective in tunnel channel. Quantum effects strongly influence the electrons flow through these devices. Coulomb gap, Coulomb staircases and clear periodic peaks are seen directly at room temperature in the current-voltage (I-V) characteristics over entire devices.

ACKNOWLEDGMENTS

The authors gratefully acknowledge support of the National Science Council of the Republic of China under the contract No. NSC94-2112-M-492-003. We are greatly indebted to the research staffs in National Nano Device Laboratories (NDL) for providing much help.Begin typing text here

REFERENCES

1. Phys. World **13**, 12 (2000); also at http://www.bell labs.com/news/1999/november/15/vertical.pdf.
2. K. Nakazato, K. Itoh, H. Mizuta, and H. Ahmed, *Electron. Lett.* **35**, 848 (1999).
3. S. Tarucha, D. G. Austing, and T. Honda, *Phys. Rev. Lett.* **77**, 3613 (1996).
4. H. Fukuda, J. L. Hoyt, M. A. McCord, and R. F. W. Pease, *Appl. Phys. Lett.* **70**, 333 (1997).
5. H. Fukuda, J. L. Hoyt, M. A. McCord, and R. F. W. Pease, *Appl. Phys. Lett.* **70**, 333 (1997).

6. D. M. Pooley, H. Ahmed, H. Mizuta and K. Nakazato, Coulomb blockade in silicon nano-pillars, *Appl. Phys. Lett.* **74**, 2191 (1999).
7. D. M. Pooley, H. Ahmed, and N. S. Lloyd, Fabrication and electron transport in multilayer silicon-insulator-silicon nanopillars, *J. Vac. Sci. Technol.* B 17., 3235 (1999).
8. D. M. Pooley, H. Ahmed, H. Mizuta and K. Nakazato, Single-electron charging phenomena in silicon nanopillars with and without silicon nitride tunnel barriers, *J. Appl. Phys.* **90**, 4772 (2001).
9. P. A. Lewis, B. W. Alphenaar, and H. Ahmed, Measurements of geometric enhancement factors for silicon nanopillar cathodes using a scanning tunneling microscope, *Appl. Phys. Lett.* **79**, 1348 (2001).
10. G. L. Ingold and Yu. V. Nazarov, in Single Charge Tunneling, edited by H. Grabert and M. Devoret (Plenum, New York, 1992).
11. T. Pohjola, Dynamical Properties of Single-electron Devices and Molecular Magnets, PhD. Thesis of Helsinki University of Technology, Finland, 2001.
12. T. H. Oosterkamp, T. Fujisawa, W. G. van der Wiel, K. Ishibashi, R. V. Hijman, S. Tarucha, and L. P. Kouwenhoven, *Nature* **395**, 873 (1998)

Mater. Res. Soc. Symp. Proc. Vol. 913 © 2006 Materials Research Society   0913-D03-04

Effect of Spacer Scaling on PMOS Transistors

Wai Shing Lau[1], Chee Wee Eng[2], David Vigar[2], Lap Chan[2], and Soh Yun Siah[2]
[1]School of EEE, Nanyang Technological University, NTU, School of EEE, Block S2.1, Nanyang Avenue, Singapore, Singapore, 639798, Singapore
[2]Chartered Semiconductor Manufacturing Ltd, Woodlands Industrial Park D St. 2, Singapore, Singapore, 738406, Singapore

ABSTRACT

Our observation is that both the on-current and off-current of state-of-the-art p-channel MOS transistors tend to become larger when the L-shaped spacer becomes smaller due to two different mechanisms: a decrease in the effective channel length Leff (Mechanism A) and a decrease in the series resistance (Mechanism B). In our analysis, we use drain induced barrier lowering (DIBL) as a measure of Leff and we assume that there is a linear relationship between the on-current, the logarithm of the off current and DIBL. Our assumption is supported by our theoretical derivations.

INTRODUCTION

As CMOS technology is scaled down to smaller dimensions, the effective channel length (Leff) of MOS transistors becomes smaller and smaller. The spacer also becomes smaller in size and gradually conventional spacer process migrates to L-shaped spacer process. During state-of-the-art CMOS integrated circuit manufacturing, we observed that the p-channel MOS (PMOS) transistor shows up a sizeable increase in the on-current and off-current when the L-shaped spacer becomes slightly smaller. The effect of a slightly smaller spacer is pretty weak for the corresponding n-channel MOS (NMOS) transistor. It is known that boron TED (transient enhanced diffusion) and BED (boron enhanced diffusion) of the p-type source/drain extension induced by the p-type deep source/drain implant have strong effects on PMOS transistors [1]. Si interstitials generated by implant damage (TED) or the presence of a large quantity of B (BED) can enhance B diffusion through the following reaction:

$$B_S + I \leftrightarrow BI ,$$ (1)

where B_S is the substitutional boron, I the Si self-interstitial and BI the interstitial-assisted mobile boron.

0.11 μm CMOS technology using L-shaped spacer technology was used to fabricate MOS transistors. For our experiment, liner spacer splits A and B have a liner spacer thickness of 8 and 12 nm, respectively. We can easily imagine that there are two possible physical mechanisms which can increase the on-current. Mechanism A is that the effective channel length becomes smaller. It can be imagined that a slightly smaller spacer reduces the distance between the p-type deep source/drain implant and the channel region, resulting in stronger boron TED (transient enhanced diffusion) and BED (boron enhanced diffusion) of the p-type source/drain extension, which can cause a reduction if the effective channel length (Leff) of the PMOS transistor. This will cause a significant increase in the on-current and off-current. Mechanism B is that the drain/source series resistance becomes smaller. It can also be imagined that a slightly smaller spacer can cause a reduction in the series resistance of the PMOS transistor. In this paper, we will investigate the physical mechanisms involved in the increase in the PMOS on-current and off-current when the spacer becomes slightly smaller.

It is a challenge to find out whether Mechanism A or Mechanism B is the dominant mechanism because there is a significant within-wafer variation and wafer-to-wafer variation in the effective channel length and the drain/source series resistance. In addition, accurate drain/source series resistance measurement is necessary in order to distinguish a small change in the drain/source series resistance. Accurate drain/source series resistance measurement can also be challenging. In this paper, we use DIBL (drain-induced-barrier-lowering) as a measure of effective channel length [2]. We propose to use a relative drain/source series resistance technique instead of an absolute drain/source series resistance technique because the former can be performed much more easily than the latter.

RESULTS & DISCUSSION

As shown in Fig. 2, on-current is plotted against DIBL for wafers using Liner Spacer Split A and Liner Spacer Split C conditions. It can be seen that the data points can be fitted by basically two parallel lines, represented by the following equations:

Liner Spacer Split A: Idsatp_um(uA/um)=1.2805DIBL+169.36   (2)

Liner Spacer Split C: Idsatp_um(uA/um)=1.2708DIBL+165.11   (3)

DIBL is in mV for equations (2) & (3). It can also be seen that DIBL (Leff) is larger (smaller) for Liner Spacer Split A than Liner Spacer Split C, as shown in Table II. From Table II, there is an increase of on-current by 367-349=18 uA/um when the process is shifted from Liner Spacer Split C to Liner Spacer Split A. The effect of the drain/source series resistance can be estimated from equations (2) & (3) to be 169.36-165.11= 4 uA/um approximately. Thus an increase of on-current by 14 uA/um can be attributed to Mechanism A (Leff decrease) while an increase of on-current by 4 uA/um can be attributed to Mechanism B (series resistance decrease). Thus Mechanism A (Leff decrease) is the more important mechanism compared to Mechanism B (series resistance decrease). We have developed an equation for a relative series resistance measurement method as follows.

$$\Delta R_{ds} = (Vd/Idsat)- (Vd/Idsat\_predicted) \qquad (4)$$

Vd is the drain voltage used for Idsat measurement. Idsat is the drain saturation current actually measured while Idsat_predicted is the drain saturation current predicted. For example, we can use data from Liner Spacer Split A as described by equation (2) to generate Idsat_predicted for

Liner Spacer Split C. Idsat comes from equation (3). Then we substitute into equation (4) the actual measured and predicted drain saturation currents for Liner Spacer Split C. The drain/source series resistance can be estimated to be about 46 Ω-um smaller for Liner Spacer Split A than Liner Spacer Split C on the average with a standard deviation of 23 Ω-um.

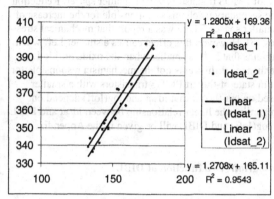

Figure 2. Idsatp_μm vs. DIBL plot for liner spacer split.

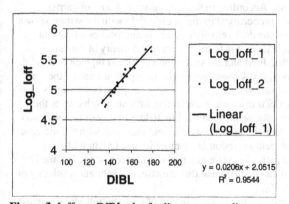

Figure 3. Ioff vs. DIBL plot for liner spacer split.

As shown in Fig. 3, off-current is plotted against DIBL for wafers using Liner Spacer Split A and Liner Spacer Split C conditions. It can be seen that the data points can be fitted by basically one line. It can also be seen that DIBL (Leff) is larger (smaller) for Liner Spacer Split A than Liner Spacer Split C.

CONCLUSIONS

In conclusion, we show that Leff reduction because of faster boron diffusion due to enhanced TED/BED resulting from a smaller spacer is the primary mechanism responsible for the increase of on-current and off-current of PMOS transistors with a smaller spacer. Reduction of the drain/source series resistance is the secondary mechanism responsible for the increase of on-current of PMOS transistors with a smaller spacer. There was a report that mechanical stress in nitride spacer enhances boron diffusion [3]. However, this effect causes a smaller Leff when the nitride spacer is bigger. Our observation is that Leff is smaller when the oxide/nitride L-shaped spacer is smaller. Hence, stress-enhanced diffusion is not the mechanism responsible for an increase in on-current and off-current in state-of-the-art PMOS transistors with a smaller spacer. In our paper, our treatment is based on our theory that Idsat and log(Ioff) depend on DIBL linearly. A more detailed theory regarding the linear relationship between Idsat and DIBL and also the linear relationship between log(Ioff) and DIBL will be given in the Appendix.

APPENDIX: Lau's theory on Idsat or log(Ioff) as a function of DIBL

A. Idsat as a linear function of DIBL

In this appendix, we will discuss our theoretical explanation of the linear relationship between the saturation current and DIBL. According to the conventional theory of carrier transport, the drift velocity of carriers is proportional to the electric field with a constant known as mobility at low electric field and then the drift velocity of carriers approaches a constant saturation velocity at high electric field. According to the conventional theory of carrier transport, the drift velocity at low electric field and the saturation velocity at high electric field are two independent quantities. However, it is known that carrier transport in state-of-the-art MOS transistors does not follow the conventional theory of carrier transport. Lundstrom's theory [4]-[5] shows that in state-of-the-art MOS transistors, the effective saturation velocity of the inversion layer at high electric field actually depends on the low field mobility of the inversion layer such that the effective saturation velocity at high electric field increases with the increase of the low field mobility. Since the low field inversion layer mobility is a function of the effective field perpendicular to the channel, which is a function of the gate voltage V_G, the low field mobility is a function of V_G and it can be imagined that the effective saturation velocity of the inversion layer is also a function of V_G.

The equation for the drain current in the saturation region of a submicron MOS transistor is as follows:

$$Id = v_{sat\_eff}WC_{ox,inv}(V_G - V_{th,sat\_IV}) \qquad (A1)$$

In equation (A1), v_{sat_eff} is the effective saturation velocity, W is the channel width, $C_{ox,inv}$ is the oxide capacitance per unit area at inversion and V_{th,sat_IV} is the threshold voltage used to create a linear fit to Id vs. V_G relationship. Our theory is that v_{sat_eff} is not a constant; instead it is a function of V_G. When V_G is small, there is strong Coulombic scattering such that v_{sat_eff} is smaller; when V_G increases, carriers generated by inversion cause a screening of Coulombic scattering centers such that v_{sat_eff} increases When V_G approaches V_{DD}, the power supply voltage , v_{sat_eff} approaches a constant value. V_{th,sat_IV} is the threshold voltage used to create a linear fit to Id vs. V_G relationship in the neighbourhood of $V_G = V_{DD}$; V_{th,sat_IV} tends to be significantly larger than saturation threshold voltage defined in the conventional manner.

Then, $$gm = v_{sat\_eff}WC_{ox,inv} \tag{A2}$$

When $V_G = V_{DD}$, Id = Idsat.

$$Idsat = v_{sat\_eff}WC_{ox,inv}(V_{DD}-V_{th,sat\_IV}) \tag{A3}$$

$$or \ Idsat/W = v_{sat\_eff}C_{ox,inv}(V_{DD}-V_{th,sat\_IV}) \tag{A4}$$

Then, $$gm\_sat = gm(V_G=V_{DD}) \tag{A5}$$

DIBL is usually measured in the sub-threshold region of MOS transistors at a certain constant current, for example, 0.1 $\mu A(W/L)$, where W is the nominal gate width and L is the nominal gate length.

$$DIBL = V_{th,lin\_CC} - V_{th,sat\_CC} \tag{A6}$$

We have tried to fit the experimental results with

$$V_{th,sat\_IV} = Q - P \ (DIBL) \tag{A7}$$

In equation (A7), Q and P are two constants.

Substitute (A7) into (A4) and re-arrange,

$$Idsat/W = P(v_{sat\_eff}C_{ox,inv})DIBL + (v_{sat\_eff}C_{ox,inv})(V_{DD}-Q) \tag{A8}$$

Then, $$Idsat/W = (v_{sat\_eff}C_{ox,inv})[P*DIBL + (V_{DD}-Q)] \tag{A9}$$

However, experimentally we observed that v_{sat_eff} is not totally constant at a constant drain voltage. It can be easily imagined that v_{sat_eff} slightly decreases when Leff increases such that DIBL decreases. Thus we can model v_{sat_eff} as follows.

$$v_{sat\_eff} = F + G \times DIBL \tag{A10}$$

In equation (A10), F and G are two constants; G is a small number such that v_{sat_eff} has a weak dependence on DIBL. Substitute equation (A10) into equation (A9) and neglect second order effect, we have

$$Idsat/W = C\_1*DIBL + C\_2 \tag{A11}$$

In equation (A11), C_1 and C_2 are two constants. Thus there is a linear relationship between Idsat and DIBL. In addition, experimental results show that $C\_1/C_{ox,inv}$ is close to the known carrier saturation velocity by a factor of 2. Further investigation is ongoing.

B. Log(Ioff) as a linear function of DIBL

Regarding the relationship between log(Ioff) and DIBL, it is possible to imagine that there is a parasitic bipolar transistor in the MOS transistor such that the off current is given by the following equation:

$$Ioff/W = I_{BJTO}exp(V_{BE\_eq}/V_T) \tag{A12}$$

In equation (A12), I_{BJTO} is a constant and V_T is equal to kT/q where k is the Boltzmann constant and q is the electronic charge. DIBL is actually an effective measure of equivalent VBE (i.e. V_{BE_eq}) applied to the parasitic bipolar transistor due to the drain voltage such that

$$V_{BE\_eq} = constant \times DIBL \tag{A13}$$

Combining equations (A12) and (A13) will yield a linear relationship between log(Ioff) and DIBL such that

$$Log(Ioff) = C\_3 \times DIBL + C\_4 \tag{A14}$$

In equation (A14), C_3 and C_4 are two constants. Thus there is a linear relationship between log(Ioff) and DIBL. Combining equations (A11) and (A14) and eliminating DIBL, it can be easily seen that there is a linear relationship between Idsat and log(Ioff), which can be observed experimentally [6].

REFERENCES

1. S. Shishiguchi, A. Mineji and T. Matsuda, Jpn. J. Appl. Phys., vol. 42, pp. 7265-7271 (2003).
2. Q. Ye and S. Biesmans, Solid-State Electron., vol. 48, pp. 163-166 (2004).
3. P. Eyben, N. Duhayon, C. Stuer, I. De Wolf, R. Rooyackers, T. Clarysse, W. Vandervorst and G. Badenes, MRS Symp. Proc, vol. 669, pp. J2.2.1-J2.2.6 (2001).
4. M. S. Lundstrom, IEEE Electron Dev. Lett., vol. 18, pp. 361-363 (1997).
5. M. S. Lundstrom, IEEE Electron Dev. Lett., vol. 22, pp. 293-295 (2001).
6. H. Liao, P.S. Lee, L.N.L. Goh, H. Liu, J.L. Sudijono, W. Elgin and C. Sanford, Thin Solid Films, vol. 462-463, pp. 29-33 (2004).

Mater. Res. Soc. Symp. Proc. Vol. 913 © 2006 Materials Research Society   0913-D03-08

Low Temperature Silicon Dioxide Deposition and Characterization

d Chatham, Martin Mogaard, Yoshi Okuyama, and Helmuth Treichel

za Technology, Inc., 440 Kings Village Road, Scotts Valley, CA, 95066

ABSTRACT

Although much effort has been expended toward developing alternate dielectrics for use in abricating ULSI circuits, there is still a need for high quality SiO$_2$ films. In particular, process emperature restrictions have increased the demand for high quality, low temperature SiO$_2$ films.[1] Such films have multiple applications in microelectronics, including use as passivation coatings, nterlevel dielectrics, gate dielectrics in metal oxide semiconductor field effect transistors (MOSFETs), hin film transistors, and in devices using dual spacers.[2] Advanced devices at the 65-nm node and beyond are typically fabricated with nickel silicided electrodes—which enable lower junction silicon consumption, lower sheet resistance, and reduced agglomeration, but require subsequent process emperatures to be below 550°C. Also, to prevent movement of the ultra shallow junctions (USJs) during a subsequent thermal cycle, the temperatures for process steps after USJ formation must be kept below 600°C. To meet these needs, we have developed a low temperature (<500°C) SiO$_2$ process that results in excellent dielectric quality.

This paper presents results on high-quality chemical vapor deposition (CVD) SiO$_2$ films deposited at temperatures from 200°C to 450°C using a novel proprietary and versatile silicon precursor using oxygen as the oxidizer. Composition, film stress, deposition rate, leakage current density, and step coverage results are presented.

INTRODUCTION

The need for low temperature SiO$_2$ is driven by continued decrease in processing temperatures. For example, process temperatures must also be kept below 600°C to prevent movement of the USJs during subsequent processing. The use of metal silicides such as NiSi in advanced transistors has imited process temperatures to below 550°C. Microelectronics applications of high quality SiO$_2$ thin films deposited at low temperatures, including: passivation coatings, interlevel dielectrics, liners and devices using dual-spacers. Low temperature CVD is one approach being explored to achieve the goal of depositing SiO$_2$ films. In this paper, we present results on the behavior of a low temperature oxide film deposited at low pressures from a carbon and chlorine-free silicon precursor.

EXPERIMENTAL METHOD

SiO$_2$ films were deposited by CVD from a novel high vapor pressure precursor reacting with oxygen. The films were deposited in a 200/300 mm single wafer ALD system operated in CVD mode over a temperature range of 200°C to 450°C using 200 mm wafers. Thickness and refractive index at 633 nm results were determined from spectroscopic ellipsometry measurements at 49 points with 3 mm edge exclusion, using a SiO$_2$ model. Film stress was determined from measurements of the change of wafer bow. As-deposited leakage current density versus voltage (I-V) was measured at 5 points using a mercury probe with 7.15×10^{-5} cm^2 contact area. Post deposition annealing was carried out in a furnace at atmospheric pressure at 700°C for 30 min in N$_2$. Wet etch rates were measured on as deposited and annealed films using 50:1 HF.

RESULTS

Results are presented from a study of the influence of deposition temperature (200°C to 450°C) o film properties. The deposition rate increases slightly with increasing temperature whereas withi wafer uniformity is insensitive to deposition temperature (see Figure 1). As deposited refractive inde declines with increasing deposition temperature (see Figure 2), approaching the index of quartz (1.445 The index after annealing is about 1.45 at all deposition temperatures.

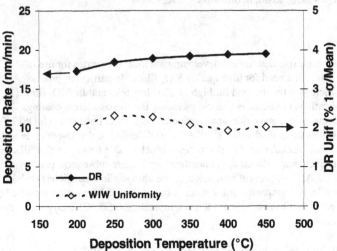

Figure 1. The dependence of the deposition Rate (DR) and within wafer uniformity (1-σ/Mean) o deposition temperature at an O_2: reactant ratio of ~20:1.

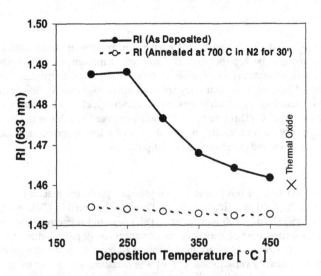

Figure 2. The refractive index at 633 nm versus temperature as deposited and post 700°C, 30 min N annealing. The thermal oxide RI value is for reference.[3]

ilm stress levels were moderate at all deposition temperatures (See Figure 3), with films compressive s deposited (with stress magnitude increasing with increasing deposition temperature). Post nnealing, the stress became tensile at low stress levels. Film stress levels are low compared to the vels required for strained devices. Film shrinkage (Figure 4) also decreases with increasing eposition temperature.

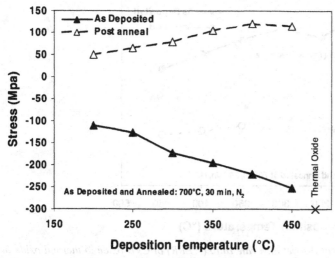

Figure 3. The dependence of low temperature CVD SiO_2 film stress on deposition temperature for as eposited and annealed films. Room temperature thermal oxide stress value from Ref. 4.

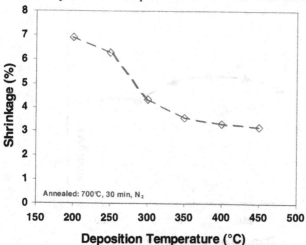

Figure 4. The dependence of film shrinkage after annealing at 700°C for 30 min in N_2 on deposition emperature.

The film wet etch rate ratio (Figure 5) for as deposited films is high (10 – 12) but drops to ~6 after annealing. The ratio drops slightly with increasing deposition temperature for as deposited and annealed films. The leakage current density at a bias of –1.5 V for ~10 nm thick films (E = ~1.5 MV/cm) is approximately independent of deposition temperature (see Figure 6).

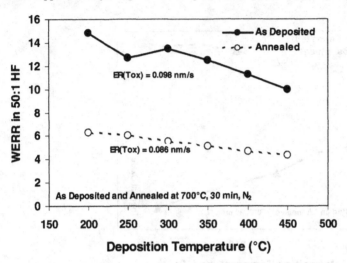

Figure 5. The dependence of the wet etch rate ratio (WERR) of CVD oxide to thermal oxide in 50:1 HF on the CVD oxide deposition temperature. Etch rates (ER) of thermal oxide with and without annealing are displayed in the graph.

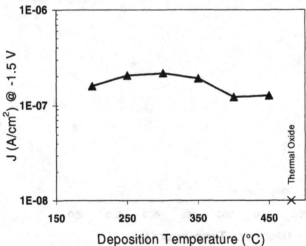

Figure 6. The dependence of the leakage current density measured at –1.5 V of as deposited ~10 nm thick CVD oxide films on the deposition temperature. Thermal oxide leakage data from reference 5.

Composition data obtained by X-ray photoelectron spectroscopy (XPS) for ~10 nm films deposited at 250°C and 450°C is shown in Table 1. Measurements made before and after Ar sputtering shows a drop in N content after sputter cleaning, suggesting some of the residual N content is on the film surface. As the film is ~10 nm thick, some Si signal from the underlying Si substrate is contributing to the Si level, resulting in O:Si levels less than 2.

Temp	As Deposited (Atomic %)				After Ar Sputter Cleaning (Atomic %)			
	Si	O	N	O:Si	Si	O	N	O:Si
200°C	33.6	62.5	1.2	1.86	37.2	62.3	0.30	1.67
450°C	36.9	64.1	2.2	1.74	36.8	62.2	1.21	1.69

Table 1. SiO₂ film composition determined from X-ray photoelectron spectroscopy. Si and O levels not corrected for Si° signal from substrate.

STEP COVERAGE RESULTS

Step coverage results were obtained on 65 nm shallow trench isolation (STI) (aspect ratio, AR, >4) and 250 nm metal 1 (AR ~ 3) structures (see Figures 7a, 7b). These figures show excellent conformality. Other tests on deep trench structures with high aspect ratios show 80% conformality at the bottom of the trench.

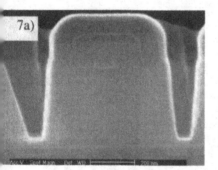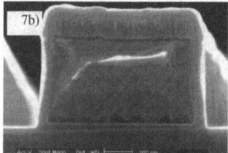

Figures 7a, b. Conformality of the CVD SiO₂ film deposited in a single wafer system at 450°C using O₂ on a) STI and b) Metal 1 structures.

CONCLUSIONS

We have demonstrated a low temperature CVD SiO₂ process using a novel carbon- and chlorine-free precursor resulting in good quality SiO₂ films with excellent conformality. Results over the temperature range 200°C to 450°C show that the deposition rate and leakage current are relatively insensitive to deposition temperature. Shrinkage and RI measurements indicate that film properties improve from 200°C to 350°C, with less change up to 450°C.

ACKNOWLEDGEMENTS
The authors thank B. Ford for providing SEM photos.

REFERENCES
1. B. Park, R. Conti, L. Economikos, A. Chakravarti, and J. Ellenberger, J. Vac. Sci. Technol. B. 19 (5), 1788-95 (2001).
2. F. S. Becker, in *Reduced Thermal Processing for VLSI*, edited by A. L. Roland (Plenum Press, New York, NATO ASI Series, Series B: Physics, Vol 207, 1989), pages 355 – 392.
3. D. Dobkin, ww.batnet.com/enigmatics/semiconductor_processing/CVD_Fundamentals/films/SiO2_Properties.html, page 3.
4. S. Wolf and R. N. Tauber, *Silicon Processing for the VLSI Era*, Vol. 1, (Lattice Press, Sunset Beach, CA, 1988), page 228.
5. T. Qiu, C. Porter, M. Mogaard, J. Bailey, and H. Chatham, submitted to *Solid State Technology*, July 2006.

Mater. Res. Soc. Symp. Proc. Vol. 913 © 2006 Materials Research Society   0913-D03-12

BCl₃/N₂ Plasma for Advanced Non-Si Gate Patterning

Denis Shamiryan[1], Vasile Paraschiv[1], Salvador Eslava-Fernandez[1,2], Marc Demand[1], Mikhail Baklanov[1], and Werner Boullart[1]

[1]AMPS, IMEC, Kapeldreef 75, Leuven, 3001, Belgium
[2]Catholic University of Leuven, Leuven, 3001, Belgium

ABSTRACT

A BN-like film can be deposited from a BCl_3/N_2 based plasma in a standard etch chamber at temperatures as low as 60°C. Deposition rate can be varied from 10 nm/min to more than 100 nm/min. The film contains hexagonal BN, but is very unlikely to be a stoichiometric BN. It decomposes at elevated temperatures and is water soluble. The latter property makes the post etch clean easy. The film can be used for sidewall passivation during the patterning of advanced non-Si gates. We are presenting as an example the use of BCl_3/N_2 plasma for patterning of Ge gates. The Ge gate profile is damaged by pure BCl_3 plasma during HfO_2 removal after the gate patterning. However, BCl_3/N_2 plasma with 10% N_2 preserve the gate profile while removing the high-k.

INTRODUCTION

As conventional materials in CMOS manufacturing (Si as a gate material and SiO_2 as a gate dielectric) approach their performance limit, the search for new materials becomes key point. Patterning of the stacks containing these new materials require both new plasma etch chemistries and new approaches.

Metal gates are seen as a replacement of conventional poly-Si gates since metals can alleviate issues inherent in Si gates such us poly depletion and boron penetration into the channel region. There is a wide variety of potential candidates for metal gates. Often their implementation requires adaptation of existing processing steps, in particular, gate patterning. In most cases the existing chemistries developed for poly-Si gate etch are not producing the required results if applied to metal gates. Investigation of new chemistries is needed.

One promising chemistry used for metal gate etch is based on BCl_3 [1]. It produces chlorine species that form volatile compounds with the etched metal; it forms passivating Si-B bonds with a silicon substrate [1]; it can etch metal oxides (in contrary to Cl_2) by formation of $BOCl_x$ [2,3] compound eliminating the need of a breakthrough step if the metal to be etched is oxidized. The two latter properties point out BCl_3 as a good candidate for high-k (which are mostly metal oxides) removal (where metal oxides should be removed selectively to Si substrate) [4].

However, in some cases BCl_3 might be too aggressive and produce profile distortion due to lateral etch of the gate. As will be shown below, we observed this distortion when BCl_3 was used for etching Ge gates. In order to preserve the gate profile, the lateral etch rate with BCl_3

should be reduced. One way to diminish this isotropic effect is to passivate side walls during the etch.

The sidewall passivation was achieved by adding N_2 to BCl_3 plasma. It is known that BCl_3/N_2 gas mixture can be used for chemical vapor deposition (CVD) [5] of plasma-enhanced CVD BN films [6]. It should be noted that elevated deposition temperatures (390°C-650°C) were used in order to obtain BN films of good quality. From the patterning perspective, BCl_3/N_2 plasma was reported to be used for etching of GaAs [7], the nitrogen addition was found to increase etch rate attributed to increased dissociation rate of BCl_3, however, no formation of BN layer or sidewall passivation were reported.

In this work we report that a BCl_3/N_2 based plasma produces a BN-like film and this can be used for sidewall passivation during etching of Ge gates. Since the etch is performed at relatively low temperatures (60°C) the BN-like film is of poor quality, in particular, non-stable and water soluble. This makes post etch clean very easy.

EXPERIMENTAL DETAILS

Before using the BCl_3/N_2 based plasma mixture for etching purposes, we studied film deposition and properties from the aforementioned plasma on blanket silicon wafers. Both etch and deposition experiments were carried out in a commercially available Lam Research Versys 2300™ etch reactor configured for 200 mm wafers. This is a transformer-coupled plasma (TCP) reactor that allows separate control of the plasma power and substrate bias. The substrate temperature during the experiments was maintained at 60°C.

The film deposition was performed at zero substrate bias on blanket Si (100) 200 mm wafers. Thickness of the resulting film was measured by a spectroscopic ellipsometer Sentech SE800. Film composition was studied by Fourier-Transform Infrared spectroscopy (FTIR) and X-ray Photoelectron Spectroscopy (XPS).

Poly-Ge gates were used for patterning experiments. The stack (from top to bottom) looks as follows: 193 nm photoresist with organic BARC, 60 nm SiO_2 hard mask, 100 nm poly-Ge, 3 nm HfO_2 on Si substrate. The gate patterning is hard mask (HM) based, i.e. the resist is stripped in the etch chamber directly after HM patterning and then followed by the etching of the rest of the gate stack. After etch the gate profiles were inspected by Scanning Electron Microscopy (SEM).

RESULTS AND DISCUSSION

The study was performed in two steps: first, characterization of film deposition from a BCl_3/N_2 based plasma and then application of this plasma for gate patterning.

We have found that plasma mixture of BCl_3 and N_2 gases deposits a film with a rate that varies as a function of deposition conditions (power, pressure, BCl_3 to N_2 ratio) from about 10 nm/min to more than 100 nm/min. The deposition rate dependence on the TCP power is shown in Figure 1(a) and dependence on N_2 content in Figure 1(b).

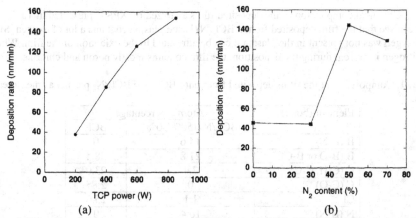

(a)   (b)

Figure 1. Film deposition rate as a function of TCP power in BCl_3/N_2 (50%/50%) plasma (a) and as a function of N_2 content in the gas mixture (b). The deposition pressure is 10 mT, TCP power is 450W (b)

The FTIR spectra of films deposited with varied BCl_3/N_2 ratio are shown in Figure 2. The prominent peak at about 1400 cm^{-1} is attributed to hexagonal BN structure (the stretching of the in-plane B-N bond).

It should be noted that the film is deposited even in the absence of nitrogen (in that case, obviously, it can't be BN). The BN peak increases with N_2 concentration but when N_2 content is too high the peak does not increase anymore indicating the fact that all boron available for this reaction is converted to BN.

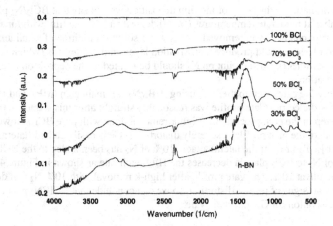

Figure 2. FTIR spectra (shifted for clarity) of films deposited from BCl_3/N_2 plasma with different gas ratios. The plasma power is 450W, pressure 10 mT.

Chemical composition of the deposited films analyzed by XPS is presented in Table I. One can see that the film deposited from a BCl_3/N_2 based plasma contains a lot of oxygen. Since the oxygen was not present in the plasma it can be attributed to the oxidation of the film. When no nitrogen is present during the deposition, the film contains mostly boron and chlorine.

Table I. Composition of the films deposited from pure BCl_3 and BCl_3/N_2 plasma as measured by XPS.

Element (bond)	Atomic percentage	
	BCl_3/N_2 (50%/50%)	BCl_3
B (B-N)	14.6	0
B (B-O or B-Cl)	21.8	38.4
C (C-H, C-C)	7.4	23.0
C (C-O)	0	2.85
Cl	1.1	25.6
N (B-N)	16.4	0
N (N-O)	3.3	3.3
O	35.5	6.9

It was found that the film is hydrophilic (contact angle for all the films is less than $15°$) and, moreover, it is water soluble. As boron nitride is insoluble in water [8], we can suggest that the deposited film is not a stoichiometric BN of good quality. The films showed instability as their thickness and refractive index changed in time. They started to decompose already at temperatures as high as $80°C$. Being a disadvantage for a deposition process, the weakness and water solubility of the film is an advantage for the etch passivation purposes: a water rinse after an etch process would be enough to remove the passivating film after etch, when it's not needed anymore.

Summarizing the study of the film deposition we can say that BCl_3/N_2 plasma run in an etch chamber at moderate temperatures ($60°C$) deposits a BN-like film with tunable deposition rates. The film can be easily removed as it is water soluble. Addition of small amounts of N_2 to a BCl_3 based plasma should produce a passivation layer on the sidewalls (which are not subjected to ion bombardment). The amount on N_2 should be limited as the deposition rate increases with N_2 concentration.

Poly-Ge gates were patterned using HBr/N_2 for main etch (ME) and HBr/O_2 to stop on HfO_2 high-k dielectric. Gate profile was reasonably straight after this sequence (Figure 3[a]). The next step in the etch process was high-k removal done with pure BCl_3. However, if such a step is applied, the gate profile is severely distorted as Ge is easily etched (laterally) by chlorine (Figure 3[b]). To prevent the lateral attack, 10% of N_2 has been added to the BCl_3 plasma. Addition of N_2 to BCl_3 plasma decreases the HfO_2 etch rate as shown in Figure 4, however, it is still acceptable at 10%. The gate profile after high-k removal with 10% N_2 and doubled removal time shows no signs of profile distortion, so we can conclude that BCl_3/N_2 plasma provides enough passivation for Ge gate side walls.

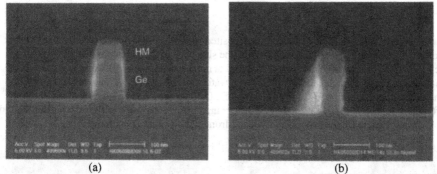

(a)   (b)

Figure 3. SEM images of Ge gate profiles, just after the gate patterning (a) and after subsequent high-k removal for 10 s with pure BCl_3 (b).

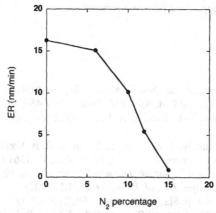

Figure 4. HfO_2 etch rate in BCl_3/N_2 plasma versus N_2 percentage. The TCP power is 450W, pressure is 5 mT, bias is 30V.

Figure 5. SEM image of Ge gate profile after high-k removal for 20 s in BCl_3/N_2 plasma (10% N_2). The plasma parameters are the same as for Figure 4.

CONCLUSIONS

A passivating BN-like film can be deposited from a BCl_3/N_2 based plasma using a standard etch reactor. This film can passivate the sidewalls of the gates preventing the lateral etch and subsequent profile distortion. The film is not a stoichiometric BN and is water soluble. This makes a post-etch clean relatively easy: a water rinse is enough to remove it completely. One of the possible applications is protection of a Ge gate from a lateral attack by BCl_3 during HfO_2 etch (high-k removal) after Ge gate patterning. Nitrogen content in the etch plasma should be kept low, otherwise the process is switched from etch to film deposition.

ACKNOWLEDGMENTS

We would like to thank Yu HongYu and IMEC Pilot line for preparation of the samples, Thierry Conard for performing the XPS measurements.

REFERENCES

1. L. Sha and J. P. Chang, *J. Vac. Sci. Technol. A*, **22**, 88 (2004)
2. D.-P. Kim, J.-W. Yeo, and C.-I. Kim, *Thin Solid Films*, **459**, 122 (2004)
3. H.-K. Kim, J. W. Bae, T.-K. Kim, K.-K. Kim, T.-Y. Seong, and I. Adesida, *J. Vac. Soc. B*, **21**, 1273 (2003)
4. K. Pelhos, V. M. Donnely, A. Kornbilt, M. L. Green, R. B. Van Dover, L. Manchanda, Y. Hu, M. Morris, and E. Bower, *J. Vac. Sci. Technol. A*, **19**, 1361 (2001)
5. T. Tai, T. Sugiyama, and T. Sugino, *Diamond. Relat. Mater.*, 12, 1117 (2003)
6. T. Sugino and T. Tai, *Jpn. J. Appl. Phys.*, **39**, L1101 (2000)
7. K. J. Nordheden, and J. F. Sia, J. Appl. Phys., 94, 2199 (2003)
8. CRC handbook of Chemistry and Physics, 85th ed. by D. R. Lide, CRC press (2004), p.4-47

Mater. Res. Soc. Symp. Proc. Vol. 913 © 2006 Materials Research Society   0913-D03-13

Layer Transfer of Hydrogen-Implanted Silicon Wafers by Thermal-Microwave Co-Activation

Y. Y. Yang[1], C. H. Huang[1], Y. -K. Hsu[1], S. -J. Jeng[1], C. -C. Tai[2], S. Lee[1], H. -W. Chen[1], Q. Gan[3], C. -S. Chu[3], J. -H. Ting[4], C. S. Lai[5], and T. -H. Lee[1,2]

[1]Dept. of Mechanical Engineering, National Central University, Chung-Li, Taiwan

[2]Inst. of Materials Science and Engineering, National Central University, Chung-Li, Taiwan

[3]United SOI Corporation, Berkeley, California, 94707

[4]National Nano Device Laboratories, Hsinchu, Taiwan

[5]Dept. of Electrical Engineering, Chang Gung University, Kwei-Shan, Taiwan

ABSTRACT

"Thermal-microwave co-activation process" is a novel thin-film transferring technology to fabricating Silicon on Insulator (SOI) material. This technology can fully transfer large-area silicon thin film onto an insulator at low temperature. In this study, the hydrogen implanted silicon substrate was irradiated by microwave at 200 degree centigrade anneal temperature to successfully achieve a completely 8" transferred layer within 5 minutes. The result of this experiment demonstrates Thermal- microwave co-activation effective to excite hydrogen ions implanted in silicon to increase not only kinetic energy but also mobility. Finally, the surface roughness of transferred layer and the quality of bonded interface were analyzed by AFM and TEM.

INTRODUCTION

After more than three decades of researches in silicon based materials and device studies, silicon on insulator (SOI) substrate become a key materials for fabricating nano-scaling IC device. In 1994, M. Bruel [1] showed a single crystal structure silicon layer transfer technique to fabricate high quality SOI materials. This technique, so-called "Smart Cut Process", includes three main steps: hydrogen ion implantation, low temperature wafer bonding, and the thermal treatment process. In Smart Cut process, silicon wafer is implanted with high dose hydrogen ions at least $5* 10^{16}/cm^2$ as a device wafer and then this device wafer is subsequently bonded with a handle wafer using low temperature wafer bonding technique to form a bonded pair. After this bonded pair is heated at moderately high temperature of 400 to 600 degree centigrade, sub-micron silicon thin film is split from device wafer and then transferred onto the handle wafer.

The expansion of the hydrogen molecular evolving from the implanted hydrogen ions interacting with silicon dangling bonds resulted in exfoliation of the silicon thin film in the heating step. The hydrogen molecules inside the micro-cavities located near the ion projected range tend to expand and merge rather than to form individual blisters due to the Ostwald ripening processing during thermal treatment. Finally, the fracture failure of ion implanted area parallel to the bonded interface near the projected ion range is formed by the sideway expansion of the cavities due to the diffusion of implanted hydrogen excited by thermal energy.

J. T. S. Lin [2] used a non-thermal method i.e. microwave irradiation at room temperature to gain the same layer transferring goal by microwave irradiation on a 4" silicon wafer. Microwave processing can lower the activity energy to speed the chemical reaction so that it leads the format of micro-cavities occurring at low temperature by directly exciting the implanted hydrogen ions and results in decreasing the critical dosage, shortening the cutting time for layer transferring. However, microwave irradiation alone at room temperature causes the formation of lots of nucleus sites of micro-voids filled by hydrogen molecule which is immobility in silicon resulting in the issue of uniformity of transferred area. This paper address the thermal-microwave co-Activation process, combined thermal effect with microwave effect, and anticipating this method could solve the issue of uniformity of transferred layer and completely transfer a silicon thin-film from the device wafer onto the handle wafer. The method of thermal-microwave co-activation to fabricate SOI substrate will be discussed in subsequent paragraph.

EXPERIMENT

This investigation compared microwave irradiation process at variety temperature and proved the advantage of exciting implanted hydrogen ion used thermal- microwave co-activation in silicon wafer. This experiment first two (100), 8-inch, p-type, 10-20 ohm-cm silicon wafers were oxidized to grow 0.5 μm silicon dioxide layer on these surface by dry-oxidation process as device wafers. After oxidation process, these device wafers were implanted by H_2^+ at a dosage of 4×10^{16} /cm^2, 160KeV (equal to 8×10^{16} /cm^2 for implanting H^+ ion at 80 KeV) to form a buried implanted H^+ layer. Then the device wafers were directly bonded with another two handle wafers at low temperature to form two bonded pairs. The result of the bonding quality was inspected by IR image to ascertain that these bonded pairs were full bonded and proceed with this experiment.

One of the two bonded pairs was treated with microwave only process at room temperature without heating during irradiation period and another one was treated with thermal-microwave co-activation process. The experimental steps and results are described: for the microwave only process: after annealing for two hours at 200 degree centigrade to enhance the bonding energy to sufficient high, the bonded pair was then irradiated by 900W, 2.45GHz microwave at room temperature and there is no other heating treatment simultaneously before layer splitting. In the thermal-microwave co-activation process, after annealing for two hours at 200 degree centigrade to enhance the bonding energy to sufficient high, the bonded pair was heated at 180 degree centigrade and kept it at this temperature to simultaneously irradiate by 900W, 2.45GHz microwave. The temperature of layer transferred process was kept at 180 degree centigrade until the layer split in the microwave-heating co-activation oven.

RESULT AND DISCUSSION

The result of layer transferred of 8-inch bonded silicon wafer pair, after using the pure microwave irradiation process for 20 minutes at room temperature, is shown in Figure 1. Some partial areas were transferred from the device wafer onto the handle wafer that proved the microwave indeed could directly excite hydrogen ion and increase the ion's kinetic energy to form the micro-voids filled by hydrogen molecule in silicon wafer and then the blisters of hydrogen molecule continuously expanded to result in the layer split along ions implanted layer. Nevertheless, the area of the transferred layer is not complete and continuous, only about fifty percent or less. The reason is that hydrogen molecule is immovably situated in silicon when it was irradiated alone by microwave at room temperature.

In the thermal-microwave co-activation process, the 8-inch bonded silicon wafer pair was heating and kept at 180 degree centigrade to enable the mobility of implanted hydrogen ions to higher. The temperature was at 180 degree centigrade to irradiate with microwave to completely transfer the 8-inch silicon thin-film from the device wafer onto the handle wafer within five minutes. The fruitful result was shown in Figure 2. Comparing with room temperature, heating at over 150 degree centigrade, the mobility of hydrogen ions in silicon can be greatly increased more than 20 times [4]. Furthermore, increasing the temperature can also improve the ability of silicon wafer for microwave absorption [5]. In addition to efficiently heighten mobility of hydrogen ions, the energy was absorbed by silicon wafer from microwave transferred to hydrogen ions in thermal-microwave co-activation process, so the hydrogen ions can gain enough kinetic energy and mobility much more than microwave process at room temperature. A great quantity of the micro-voids filled by hydrogen molecule was formed and rapidly gathered to become blisters continuously expanded and then the silicon layer was completely and uniformly cut from device wafer in a short time.

Figure 3 shows the AFM image of the top surface of transferred layer which was exfoliated by thermal-microwave co-activation process, the roughness of the scanned area was 1.295nm and it was closed to the smoothness which was split by Smart-Cut process (1.471nm). Figure 4 shows the TEM picture of cross-sectional of the SOI substrate which was fabricated by thermal-microwave co-activation process: a silicon thin-film was transferred to the top of oxide layer of the handle wafer and the quality of bonded interface was flat and neat.

CONCLUSION

Thermal-microwave co-activation process successfully solved the issue of uniformity and continuity of transferred area and completely fabricated a large dimensions SOI substrate at sufficient low temperature in a short time. It may replace the heating step of the hydrogen-implantation based layer transfer technique for dissimilar materials layer transfer with less the thermal stress generation such as GaAs on silicon which are difficult to fabricate at high temperature due to big gap of thermal expansion coefficients but silicon substrate is "transparent" for microwave and this characteristic material properties can keep away from microwave energy absorption and avoid temperature rapidly increase.

REFERENCE

[1] M. Bruel, "Silicon on insulator material technology" Electron. Lett., Vol. 31, No. 14, pp.1201 (1995)
[2] T. -H. Lee, "Semiconductor thin film transfer by wafer bonding and advanced ion implantation layer splitting technologies", Duck University, pp. 100-121 (1998)
[3] Jason T. S. Lin et al., 2002 IEEE International SOI Conference (2002)
[4] R. H. Doremus, "Diffusion of Reactive Molecules in Solids and Melts" Artech nouse, Inc., pp. 9-17 (2000)
[5] David E. C. et al., "Microwave Processing of Materials" Annu. Rev. Mater. Sci. pp. 26:299-331 (1996)

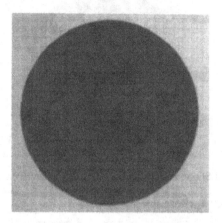

Figure 1: the photograph of 8-inch SOI material fabricated by microwave only process.

Figure 2: the photograph of 8-inch SOI material fabricated by thermal-microwave co-activation process.

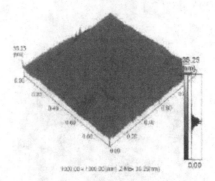

Figure 3: AFM images of SOI materials fabricated by thermal-microwave co-activation process.

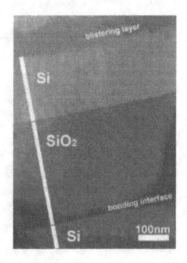

Figure 4: Cross-section TEM images of SOI materials fabricated by thermal-microwave co-activation process.

Characterization of New Materials and Structures

Mater. Res. Soc. Symp. Proc. Vol. 913 © 2006 Materials Research Society   0913-D04-01

Introduction of Airgap Deeptrench Isolation in STI Module for High Speed SiGe : C BiCMOS Technology

Eddy Kunnen, Li Jen Choi, Stefaan Van Huylenbroeck, Andreas Piontec, Frank Vleugels, Tania Dupont, Katia Devriendt, Xiaoping Shi, Serge Vanhaelemeersch, and Stefaan Decoutere
SPDT, IMEC, Kapeldreef 75, Leuven, Belgium, 3000, Belgium

ABSTRACT

The impact of capacitive coupling effects increases with scaling down the dimensions and towards higher performances. For bipolar technologies, the introduction of deep trench isolation gives a substantial reduction in the collector substrate capacitance. In this paper a method for the formation of airgap deep trenches (width $1\mu m$ – depth 6 μm) is presented. The method is fully compatible with standard CMOS Shallow Trench Isolation (STI) and does not require additional masking steps. The approach is based on a partial removal of the poly-Si filling in the trench. Subsequently, inside D-shape oxide spacers are formed narrowing the opening of the trench down. An SF_6 plasma is used to convert the nearly completely incorporated poly-Si to volatile $SiF4$, such that it desorbs through the opening. In the following steps the opening is sealed by depositing $SiO2$ resulting in the formation of an airgap (patent pending). The normal module for STI formation continues without any adaptation of the process steps. In total four standard additional process steps are needed.

The absence of the common oxide/poly filling in the deep trench decreases the peripheral collector substrate capacitance with an order of magnitude to a value of $0.02fF/\mu m$. As a consequence the low power available bandwidth is improved with 90%.

INTRODUCTION

The classical deep trench isolation has trenches filled with a poly/oxide combination. The oxide is used to lower the capacitive coupling. Complete fill with oxide is difficult because of strain issues. Having no material in the trench combines both advantages. Hence, in this paper a method for formation of deep trench airgap isolation is described. The method is compatible with STI processing since the formation occurs before the STI isolation and is compatible with CMOS processing. The subsequent CMOS processing is targeting the C013 node and has a physical gate length of 65 nm. The bipolar device has a Quasi Self Aligned structure (QSA) and within the process flow it is positioned after the CMOS processing.

THE AIRGAP DEEP TRENCH MODULE

The approach discussed starts after the formation of the buried layer formation and the subsequent epitaxial growth. In a first step a oxide-nitride-oxide hard mask is deposited on the Si. The bottom oxide avoids any strain induction of the nitride on the active areas, while the nitride itself will act as a CMP stopping layer. The top oxide layer is used as a hard mask during deep trench patterning. The trenches have a CD of 1µm and a depth into the Si of 6 µm. The trenches are etched during a single step etch in a Lam TCP9400DFM etcher with SF_6/O_2 mixture. The module continues with a channel stop implant and a reoxidation of the trench side wall. A conformal oxide liner (TEOS) is deposited followed with an α-Si and polysilicon deposition. The filling capability of amorphous silicon overcomes polysilicon filling. On the other hand the deposition rate of amorphous silicon is lower and therefore a combination of both is used. A Chemical Mechanical Polishing (CMP) step, selective to the oxide liner, planarizes the polysilicon. Because of the small trench density, good uniformity can be achieved. An anisotropic dry etch step is carried out to recess the polysilicon in the trench. Figure 1 shows a schematic and a Scanning Electron Microscope (SEM) picture of the trench at this stage of the processing.

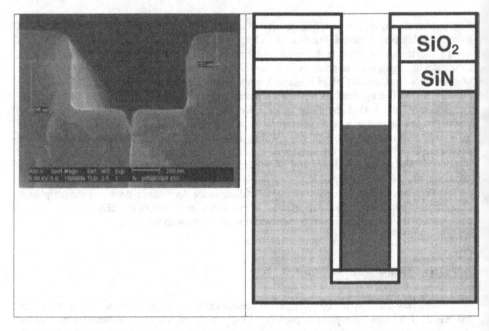

Fig 1. A schematic as well as a SEM picture after etch back of the trench, resulting in a partial filled trench. The remaining Silicon will be removed in the next steps leaving an air gap behind.

At the centre of the trench a small void appears. The void is related to little bowing of the trench in combination with a non perfect filling. The dry etch back is based upon pure HBr on a Lam Versys 2300. This chemistry allows to etch straight down, leaving no spacers behind at the sides an avoiding further aggravation of the void. Isotropic chemistries such as SF_6 would open up the void further and cause flakes. These flakes are related to deposition of material in the void that remains after removal of the polysilicon in the trench.

After recessing the polysilicon in the trench a nitride liner with an conformal oxide layer (TEOS furnace) is deposited for spacer formation. During etch, the nitride influences the endpoint signal an triggers the end of the etch. The thickness of the spacer is such chosen that a small opening remains at the centre of the trench. After spacer etching (Lam EXELAN with Carbon Fluor mixtures), the wafer is in-situ moved to the TCP9400DFM and an SF_6/HBr step without bottom power selectively removes the polysilicon out of the trench through the narrow opening between the spacers. Since dry etch is based upon formation of volatile products, the gas can enter an leave easily through the opening. Secondly, the etching of silicon with SF_6 (without O_2), does not require ions allowing an isotropic removal. A schematic of the situation after spacer formation as well as SEM inspections before and after polysilicon removal are shown in figure 2.

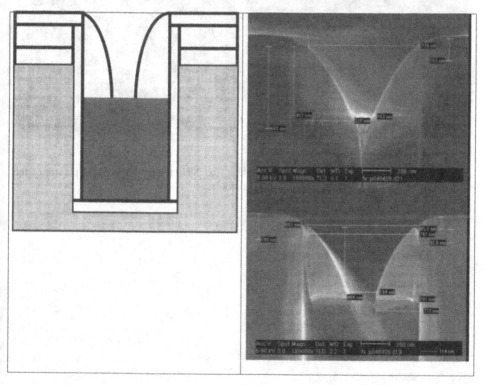

Figure 2 : A schematic presentation of the spacer formation inside the trench as well as a SEM evaluation after spacer formation and after subsequent poly removal through the opening between the spacers.

After removal, the processing continues with closing the trench. A conformal oxide deposition (furnace TEOS) closes the trench and an airgap is formed. The deposition is carried out in two steps with a intermediate anneal step. The anneal step is needed to densify the oxide layer. If not carried out, cracks appear in further processing of the STI module because of shrinkage of the oxide layer during the STI liner oxidation.

After airgap formation a CMP step is carried followed by a wet nitride removal. At this stage of the processing there is nearly no topography present and airgap deep trenches are buried into the silicon. The standard STI module can be carried out. With an appropriate chosen height of the oxide cork, the airgap remains closed during the STI and subsequent CMOS processing and bipolar processing. Figure 3 shows a SEM picture after STI etching and at the end of the STI module. Finally, the situation with back end of line processing is shown in figure 4. The approach is patent pending.

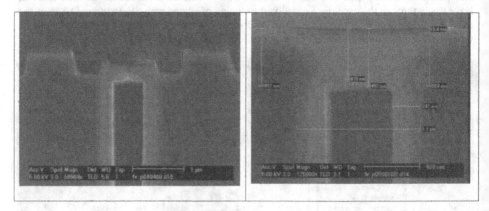

Figure 3 : Left : SEM picture of the airgap taken after STI etching. Right SEM picture of the airgap taken after the complete subsequent STI module.

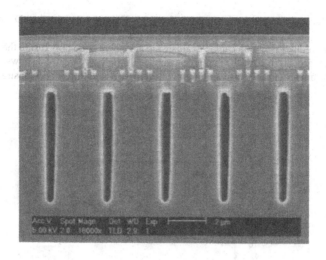

Figure 4 : Airgap deep trench isolation integrated in BiCMOS processing.

PERFORMANCE OF THE AIRGAP DEEP TRENCH ISOLATION

In this paper we focus on the performance of the Airgap Deep Trench Isolation (ADTI). An elaborated study on it's impact on device performance will be published elsewhere [VLSI]. Airgap Deep Trench Isolation is very interesting from capacitive coupling point of view. For bipolar devices, reduction can be expected in the collector substrate capacitance (C_{CS}). The C_{CS} consists of two parts : a area contribution $C_{CS,A}$ and a peripheral contribution $C_{CS,P}$. The peripheral component is related to the ADTI and becomes more important with scaling down the area. The contribution of the $C_{CS,P}$ has been limited in the past by increasing the oxide/poly ratio during filling of the trench. We refer to an oxide/poly filling when mentioning a classical DTI. A comparision of the peripheral collector substrate capacitance between STI and junction isolation, classical DTI and ADTI is presented in figure 5.

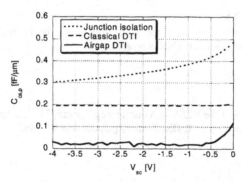

Figure 5 : Comparision of the peripheral collector substrate capacitance for Junction isolation, classical poly/oxide Deep Trench Isolation and Airgap Deep Trench Isolation.

The $C_{CS,P}$ is reduced by an order of magnitude from 0.2 fF/μm for the classical DTI towards 0.02 fF/μm for ADTI.

Secondly, high voltages isolation can be evaluated. Since there is no poly inside the trench, the possible channel around the trench cannot be switched on [PART]. Figure 6 shows the performance with respect to leakage and high voltage between collector and substrate.

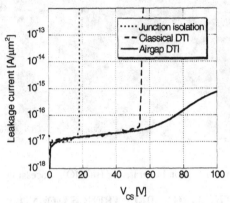

Figure 6 : Even at voltages up to 100V the leakage path around the trench remains limited because of the absence of a poly gate inside the trench.

CONCLUSIONS

An approach for making Airgap Deep Trench Isolation, compatible with STI, CMOS an bipolar processing is presented. The ADTI outperforms the classical DTI with respect to collector substrate capacitance and towards high voltage isolation. The peripheral collector substrate capacitance is reduced to 0.02 fF/μm and leakage around the trench remains below 10^{-15} A/μm^2 at 100V collector substrate voltage.

ACKNOWLEDGEMENTS

The authors would like to thank the Imec Pilot Line, the Flemish IWT and Philips Research Leuven for supporting the BiCMOS program.

REFERENCES

1. [VSLI] Proceedings of the VLSI-TSA conference Taiwan, 24 – 26 April 2006
2. [PART] Parthasarathy et al., IEDM Tech. Dig., 2002, pp. 459-462

Mater. Res. Soc. Symp. Proc. Vol. 913 © 2006 Materials Research Society   0913-D04-05

Low Temperature Selective Si and Si-Based Alloy Epitaxy for Advanced Transistor Applications

Yihwan Kim, Ali Zojaji, Zhiyuan Ye, Andrew Lam, Nicholas Dalida, Errol Sanchez, and Satheesh Kuppurao
Epi KPU, Front End Product Group, Applied Materials, 974 E. Arques Ave., M/S 81288, Sunnyvale, California, 94085

ABSTRACT

We have developed low temperature selective Si and Si-based alloy (SiGe and Si:C) epitaxy processes for advanced transistor fabrications. By lowering epitaxy process temperature (≤ 700 °C), we have demonstrated elevated source/drain formation on ultra-thin (< 50 Å) body SOI transistors without Si agglomeration, smooth morphology of selective SiGe epitaxy with high [Ge] (>30 %) and [B] (>2E20 cm^{-3}) concentrations, and selective Si:C epitaxy with high substitutional C concentration (>1 %). Also, we have increased growth rate of low temperature selective epitaxy processes by optimizing process parameters by adapting non-conventional deposition method.

INTRODUCTION

Conventional scaling metal-oxide-semiconductor (MOS) transistors for increasing device speed faces limitations beyond 65 nm technology node. In order to boost the device performance further, selective Si and Si-based alloy epitaxy processes have been recognized as solutions. Selective Si, SiGe, and Si:C epitaxy can be used for so-called elevated and recessed source/drain (S/D) applications. The elevated S/D decreases source/drain sheet resistance and also plays role as a buffer layer to protect S/D from etch damage during contact hole formation [1]. Also, elevated S/D is required to form fully depleted SOI transistors [2]. In recessed S/D application, an area below the level of the gate dielectric is etched and filled with SiGe or Si:C epitaxy depending on PMOS or NMOS transistor. This application enables one to fabricate faster transistors without shrinking the gate length by enhancing carrier mobility via uniaxial compressive or tensile strain of the transistor channel [3-6].

Need for low temperature epitaxy process results from two aspects: One is thermal budget of advanced transistor fabrication. At lower epitaxy temperatures, low growth rates are concern. The other is from a material property aspect. Strain relaxation of SiGe epitaxy and high substitutional carbon concentration in Si:C epitaxy are good examples of why epitaxy temperature should be kept low. Because the solid solubility of substitutional carbon in silicon is only $\sim 10^{17}$ cm^{-3}, carbon atoms at concentrations higher than this easily incorporate into interstitial sites or precipitate as β-SiC, resulting in no strain in the epilayer. Therefore, in order to achieve high substitutional carbon concentration ($C_{sub} \geq 1$%), non-equilibrium epitaxial growth conditions such as low temperature are required. It is well known that strained SiGe epitaxy with higher [Ge] is more easily relaxed at a given condition.

We have developed low temperature selective Si, SiGe, and Si:C epitaxy processes for advanced transistor applications and will review result of each epitaxy process.

EXPERIMENT

Selective epitaxial layers were deposited by reduced pressure thermal chemical vapor deposition in the Applied Materials Epi Centura. During selective silicon epitaxial process on a substrate patterned with dielectric films, deposition of single-crystal silicon takes place only on the exposed substrate areas but no film nucleation or growth occurs on the dielectric areas. Conventionally, selective deposition is achieved and controlled by mixing a deposition source (e.g. SiH_4) with an etchant gas such as HCl where film growth and film etching/suppression occur simultaneously. Here, we have developed an alternative approach where the growth and etching reactions are separated and optimized individually.

Selectivity and surface morphology of epitaxial layers on patterned wafers were investigated by scanning electron microscopy (SEM) and transmission electron microscopy (TEM). Ge, B, and C concentration depth profiles in SiGe and Si:C epitaxial layers were determined by secondary ion mass spectroscopy (SIMS) using Cs primary ions. High resolution X-ray diffraction (Bede Metrix L) determines strains of SiGe and Si:C epitaxy, which were also used for calculating Ge concentration ([Ge]) and substitutional carbon concentration ([Cs]) in the epitaxial layers.

RESULT and DISCUSSION

Selective Si epitaxy for elevated S/D application

Selective Si epitaxy is being used to form elevated S/D on transistor active area. It decreases sheet resistance of the S/D area and prevents it from being damaged during contact hole formation. As technology node is getting smaller, the epitaxy temperature should be reduced while the current process temperature is 800-850 °C. Also, low temperature selective epitaxy is required for fully depleted SOI application. Selective Si epitaxy processes with different deposition temperatures have been demonstrated on ultra thin (< 50 Å) silicon on insulator (SOI) patterned wafers. Figure 1 shows SEM images of the epitaxy grown at T=800 °C

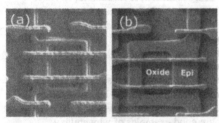
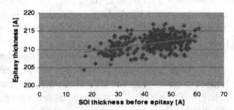

Figure 1. SEM micrographs of selective Si epitaxy grown on ultra thin (50 Å) SOI patterned wafer (a) at T=800 °C and (b) at T=700 °C.

Figure 2. Selective Si epitaxy thickness as a function of SOI thickness before the epitaxy. The epitaxy was grown at T=700 °C.

and T=700 °C. The 800 °C process results a loss of Si on S/D of the patterned wafer. It is caused by silicon migration accelerated by the higher temperature [7]. As shown in figure 1(b), however, 200 Å thick selective Si epitaxy at T=700 °C shows smooth morphology without losing the active Si area. Figure 2 shows the epitaxy thickness as a function of SOI thickness of the patterned wafer. It shows a slight increase of epitaxy thickness with SOI thickness prior to the epitaxy. Variation of the epitaxy thickness is just 15 Å with range of SOI thickness between 17 Å and 70 Å. Standard deviation of the measured epitaxy thickness data is 1% of average thickness (=210 Å). It is noteworthy that 200 Å thick Si epitaxy can be grown on 17 Å SOI layer.

Generally, growth rate of selective Si epitaxy at T ≤ 750 °C using the conventional approach is too low to be production worthy. We have developed an alternative selective process in order to realize a selective process at high deposition rate at that temperature regime. The key feature of this new process presented here is independent optimization of film growth and film etching. The selective deposition is made possible by different nucleation rates and mechanisms between on silicon crystal substrate and on dielectric film. Silicon nuclei formation on the dielectric surface is suppressed below their critical sizes by modulating the deposition and etchant gases, gas flow distribution, substrate temperature, and reactor pressure. By optimizing these process parameters, we were able to maintain a balance between epitaxial growth on silicon surfaces and no nucleation on dielectric films, which results in a high selective growth rate without losing selectivity. Figure 3 depicts cross section SEM images of selective Si epitaxy films grown at 750 °C and 700 °C on internally prepared patterned structures where growth rates are roughly a factor of three compared to those obtained by the conventional approach. However, the deposited films exhibit a minor faceting profile at corners and thus further development is needed.

 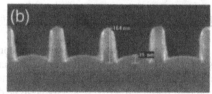

Figure 3. Cross section SEM micrographs of pattern structures processed with selective silicon epitaxy (a) at T=750 °C and (b) at T=700 °C.

SiGe epitaxy for PMOS recessed S/D application

For recent three years selective SiGe epitaxy in recessed S/D area has attracted huge attention as a way of forming compressively strained channel of PMOS transistor. In order to achieve higher compressive strain in the channel, SiGe epitaxy with higher [Ge] is required. However, increasing [Ge] in the film could result in strain relaxation and rough surface morphology. One way to avoid it kinetically is to use low temperature process. We have developed a B-doped SiGe epitaxy process at T=625 °C. The process results in [Ge]= 32 % and [B]=2.7E20 cm^{-3}, as determined by SIMS measurement (not shown here). [Ge] in SiGe epitaxy can also be determined by XRD. Figure 4 depicts XRD spectra, indicating 31.4 % of [Ge] concentration and well defined thickness fringes. From the comparison [Ge] values determined by SIMS and XRD, it can be said that there is zero or negligible strain relaxation in developed

SiGe epitaxy. The developed epitaxy process was demonstrated on patterned structure wafer. Figure 5 shows cross sectional and tilted view SEM images of 1000 Å thick SiGe epitaxy. They show smooth morphology and 100% selectivity of the epitaxy, which also supports no strain relaxation.

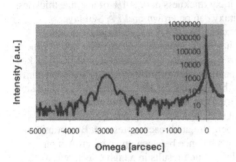

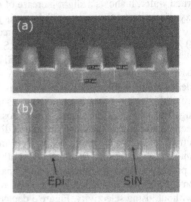

Figure 4. High resolution (004) X-ray diffraction result of selective SiGe epitaxy, which shows [Ge]=31.4% and 610 Å thickness.

Figure 5. SEM micrographs of selective SiGe epitaxy grown on patterned structure wafer. It shows 1000 Å thick (700 Å recessed + 300 Å elevated) smooth epitaxy film.

Si:C epitaxy for NMOS recessed S/D application

One of the main challenges with Si:C epitaxy is increasing substitutional carbon concentration. It is well known that lower temperature is favorable for carbon atoms to be incorporated at substitutional sites [8]. In order to have [Cs] > 1 %, the epitaxy temperature should be < 650 °C. Also, [Cs] was found to be dependent on deposition pressure. Figure 6 shows high resolution XRD spectra of selective Si:C epitaxy films grown at different pressure.

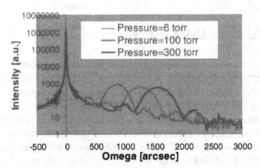

Figure 6. High resolution (004) X-ray diffraction results of selective Si:C epitaxy films grown at different pressure.

Each condition maintains the same flow rate ratio of Si source gas to carbon source gas. The epitaxy grown at higher pressure shows a clear shift of Si:C peak to higher omega angle, which means more substitutional carbon in the epitaxy. [Cs] increases to 1.6 % from 0.9 % with higher pressure. A 1000 arcsec distance of omega angle between the Si:C XRD peak and Si substrate peak indicates 1% substitutional carbon concentration using a Vegard-like linear interpolation between the lattice parameter of Si (a_{si}=5.43105 Å) and the one of cubic SiC (a_{sic}=4.35965 Å).

The optimized selective Si:C epitaxy process fills 500 Å deep recessed area. The grown epitaxy shows smooth surface morphology with 100 % selectivity, as shown in figure 7. Without optimizing process conditions, however, selectively grown Si:C epitaxial layer easily showed pitted and rough surface morphology.

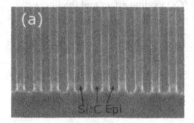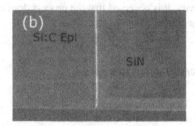

Figure 7. SEM micrographs of selective Si:C epitaxy films grown on (a) small openings and (b) on large opening of patterned structures.

Figure 8 compares two XRD spectra of selective Si:C epitaxy films grown on bare wafer and on patterned structure wafer. The comparison shows no shift of Si:C peak position, which indicates no loading effect of the carbon concentration. Also, the developed selective process showed the same growth rate regardless of Si opening size of the patterned wafer.

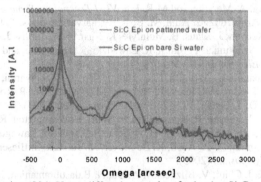

Figure 8. High resolution (004) X-ray diffraction results of selective Si:C epitaxy films grown on bare wafer and patterned structure wafer.

CONCLUSIONS

We have developed low temperature selective Si, SiGe, and Si:C epitaxy processes and demonstrated on patterned structure wafers. By reducing growth temperature of epitaxy process, elevated source/drain can be formed on ultra thin (< 50 Å) SOI patterned wafers without Si loss on active area. Also, by adapting non-conventional approach that enables us to control deposition and etch independently, we have increased growth rate (by a factor of three) of low temperature selective Si epitaxy. The developed selective SiGe epitaxy at low temperature (625 °C) shows smooth morphology with [Ge]= 32 %, [B]= 2.7E20 cm^{-3}, and 1000 Å thickness. Also, the developed low temperature selective Si:C epitaxy with [Cs]$\geq$ 1% shows smooth morphology and selectivity. It is observed that the process does not show [Cs] loading effect and thickness loading effect.

ACKNOWLEDGMENTS

The authors would like to thank Dr. W. Maszara, AMD for providing data of Figure 2 and ultra thin SOI patterned wafers.

REFERENCES

1. A. Hokazono, K. Ohuchi, K. Miyano, I. Mizushima, Y. Tsunashima, and Y. Toyoshima, *International Electron Device Meeting* (2000).
2. Z. Krivokapic, W. Maszara, F. Arasnia, E. Paton, Y. Kim, L. Washington, E. Zhao, J. Chan, J. Zhang, A. Marathe, and M-R. Lin, *VLSI Technology Symposium* (2003).
3. T. Ghani, M. Armstrong, C. Auth, M. Bost, P. Charvat, G. Glass, T. Hoffmann, K. Johnson, C. Kenyon, J. Klaus, B. McIntyre, K. Mistry, A. Murthy, J. Sandford, M. Silberstein, S. Sivakumar, P. Smith, K. Zawadzki, S. Thompson, and M. Bohr, *International Electron Device Meeting* (2003).
4. P. R. Chidambaram, B. A. Smith, L. H. Hall, H. Bu, S. Chakravarthi, Y. Kim, A. V. Samoilov, A. T. Kim, P. J. Jones, R. B. Irwin, M. J. Kim, A. L. P. Rotondaro, C. F. Machala, and D. T. Grider, *VLSI Technology Symposium* (2004).
5. G. Eneman, P. Verheyen, R. Rooyackers, F. Nouri, L. Washington, R. Degraeve, B. Kaczer, V. Moroz, A. De Keersgieter, R. Schreutelkamp, M. Kawaguchi, Y. Kim, A. Samoilov, L. Smith, P. P. Absil, K. De Meyer, M. Jurczak, S. Biesemans, *VLSI Technology Symposium* (2005).
6. K. W. Ang, K. J. Chiui, V. Bliznetsov, A. Du, N. Balasubramanian, M. F. Li, G. Samudra, and Y.-C. Yeo, *International Electron Device Meeting* (2004); Appl. Phys. Lett. **86**, 093102 (2005).
7. T. Sato, K. Mitsutake, I. Mizushima, and Y. Tsunashima, Jpn. J. Appl. Phys. **39**, 9A, pp. 5033-5038 (2000).
8. V. Loup, J. M. Hartmann, G. Rolland, P. Holliger, F. Laugier, and M. N. Semeria, J. Vac. Sci. Technol. **B 21**, 246 (2003).

Mater. Res. Soc. Symp. Proc. Vol. 913 © 2006 Materials Research Society 0913-D04-07

Nano-Scale MOSFET Devices Fabricated Using a Novel Carbon-Nanotube-Based Lithography

Jaber Derakhshandeh[1], Yaser Abdi[1,2], Shams Mohajerzadeh[1], Mohammad Beikahmadi[1], Ashkan Behnam[1], Ezatollah Arzi[2], Michael D. Robertson[3], and C. J. Bennett[3]

[1]Electrical & Computer Eng., Thin Film Laboratory, Department of Electrical and Computer Eng, University of Tehran, Tehran, Iran, Tehran, Tehran, 14395/515, Iran
[2]Department of Physics, University of Tehran, Tehran, Iran, Tehran, Tehran, 14395/515, Iran
[3]Department of Physics, Acadia University, Wolfville, Nova Scotia, B4P 2R6, Canada

ABSTRACT:

PECVD-grown carbon nanotubes on (100) silicon substrates have been studied and exploited for electron emission applications. The growth of CNT's is achieved by a mixture of hydrogen and acetylene gases in a CVD reactor and a 2-5nm thick nickel is used as the seed for the growth. The presence of DC-plasma yields a vertical growth and allows deposition at temperatures below 650°C. The grown nano-tubes are encapsulated by means of an insulating TiO_2 layer, leading to beam-shape emission of electrons from the cathode towards the opposite anode electrode. The electron emission occurs using an anode-cathode voltage of 100 V with ability of direct writing on a photo-resist coated substrates. Straight lines with widths between 50 and 200nm have been successfully drawn. Scanning electron and transmission electron microscopy have been used to investigate the quality and fineness of the results. This technique has been applied on P-type (100) silicon substrates for the formation of the gate region of N-MOSFET devices, showing a drive current of $310\mu A/\mu m$ and Cox of $0.7\mu F/cm^2$.

INTRODUCTION

Modern MOSFET devices lend their success to the fine patterning of their gate constituent. The formation of ultra-fine gate regions requires high resolution lithography techniques beyond the capability of standard photo-lithography. There are several techniques for submicron and nano-lithography among which one can refer to the electron beam writing, ion beam lithography and extreme deep ultra violet lithography [1-3]. Electron beam writing is essentially capable of generating patterns in nano-meter scale, especially down to 10 nm regime. This technique can be used for writing the gate pattern and it is a time consuming method. By scanning the electron beam over an electro-resist coated sample desired patterns are generated on the sample. Usually electron energies in the range of 10-20 keV are exploited. High density polymers as PMMA are being used to minimize the proximity effect, a side-effect of the leaching of energetic electrons.

On the other hand carbon nanotubes are suitable choices for emission of electrons, provided a proper electron beam is formed. The vertical growth of carbon nanotubes on silicon substrates is the most crucial step in achieving suitable electron emission. Several techniques are being exploited for the growth of single-wall and multi-wall nanotubes among which one can highlight the thermal catalyst-based CVD, laser vaporization deposition and arc discharge growth. These techniques, however, fail to carry out a vertical growth of CNT's on silicon substrates at low temperatures [4-6]. Plasma enhance chemical vapor deposition seems a favorite method to grow nanotubes in a vertical arrangement [7-8].

We have recently reported the vertical growth of carbon nano-tubes using a PECVD technique where a mixture of acetylene and hydrogen was used to carry the main constituents for the deposition of CNT's[10-12]. Following the growth of the tubes, they are coated with an insulating layer and the encapsulated tubes are then polished to open up the top side of the tubes. The physical characteristics of the tubes, as well as their application for sub-micron and nano-lithography are addressed. Also, this novel approach has been used to form lines with 100nm width suitable for the gate of nano-metric MOSFET devices. The electrical characteristics of the transistors realized will be presented. The lithography technique proves to be a suitable approach for making transistors at sub-micron and 100 nm scale. In the following sections we present the CNT growth, followed by the preliminary results for lithography. Finally the MOSFET fabrication is described and electrical characterization of sub-micron transistors will be presented.

EXPERIMENTAL SET-UP

The growth of carbon nano-tubes is the heart for the subsequent lithography step and it is achieved using a direct-current plasma enhanced chemical vapor deposition chamber. The DC plasma is generated between the specimen holding negative (cathode) and the opposite anode plates, respectively. Usually 1"-square silicon substrates are cleaned using a standard RCA #1 solution, followed by rinsing in D.I. water and blow-drying. The cleaned samples are placed in an electron beam evaporation system to deposit a thin layer of nickel with a thickness of 30-80Å. The Ni-coated samples are then transferred into the quartz reactor and after proper evacuation; they are heated to a temperature of 550°C for out-gassing.

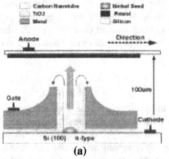

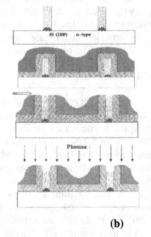

Figure 1: The diagram showing the overall operation of the self-defined electron emission sources and their fabrication process flow. The emission of electrons from the negatively biased cathode will end up with tracks of electrons on the oppositely placed, resist-coated substrate(a). And (b), the fabrication steps.

(a)

(b)

The first step in growing carbon nano-tubes is the formation of nano-sized nickel islands which is accomplished using hydrogenation of the sample with a gas flow of 30sccm of H_2 and at a pressure of 1torr. The power density for this step is set at 4W/cm^2. Immediately after this step, the growth of the carbon nanotube is achieved by introducing a mixture of H_2 and C_2H_2 (acetylene) with a ratio of 6:1 and at a pressure of 1.5torr. The temperature during this step is maintained at 650°C and the plasma power is kept at 5.5W/cm^2. The grown CNT's are encapsulated using chemical vapor deposition of titanium-oxide (TiO_2). More details about the processing of the CNT and their application for nanolithography are found elsewhere [12]. Scanning electron microscopy and transmission electron microscopy are used to observe the grown CNT's.

RESULTS AND DISCUSSIONS

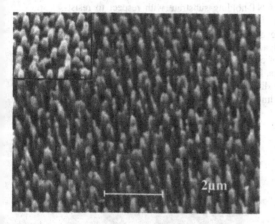

Figure 2: SEM images of a sample grown using this PECVD method. Inset shows the fully processed nanotubes with the titania encapsulating oxide.

Figure 2 shows the growth of nanotubes with a plasma power of $4.5W/cm^2$ evidencing a sparse deposition. The increase in the plasma power leads to the formation of a heavy growth in a vertical fashion. Excessive increase in the plasma power will lead to damage in the substrate and the deposited film becomes partially etched off the surface. The nanotubes grown for this study are vertical and the grown layers must be encapsulated to form a proper electron emission.

Figure 3 shows the TEM images of a single nanotube before any further processing has been performed. The figure on right-side corresponds to the sample which has been fully prepared as stated in the previous section. The individual CNT has been coated with titanium oxide, polished and ashed in oxygen plasma, which can be better conceived by referring to Fig.1.b.

Figure 3: TEM images of CNTs for this study.

Each encapsulated nanotube in this fashion can act as the source of electron emission and its trace is printed on the opposite substrate. Figure 4.a corresponds to a case where a cluster of electron emitting tips are active and Figure 4.b corresponds to the case with one individual tip acting as electron emitter. This figure demonstrates some of the resulted lines drawn with this technique. As can be seen in Fig. 4.a, parallel lines can be created on the opposite substrate as a result of electron emission combined with appropriate stage movement. For this image, a sample with a cluster of electron-emitting sources (CNT's here) has been used. Generation of straight parallel lines on the resist coated substrate has been achieved by a linear moving of the CNT-holding substrate with respect to resist-coated substrate. Fig.4 shows the evolution of straight lines drawn with a width of 100 to 200 nm and their length limited by the span that motorized stage would carry out. By applying an isolated carbon nano-tube instead of clusters, individual lines can be drawn which are most suitable for the formation of electronic devices. The line can then be used to define the gate of the MOSFET transistors. Part (b) of this image corresponds to the case where the emission of electrons from an individual CNT (shown in the inset of Fig.4.b) leads to the formation of an isolated narrow line.

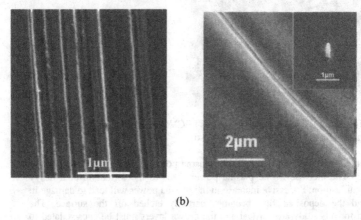

(a)   (b)

Figure 4: The SEM images corresponding to the lines drawn using the encapsulated carbon nanotubes on the resist coated substrates. (a) Main substrate is fully covered with activated carbon nano-tubes, evidencing parallel lines in all places, (b) single lines drawn corresponding to isolated nano-tubes. This latter case is most suitable for nano-lithography.

The formation of MOSFET transistors in this paper requires both photolithography for the definition of the gate, source and drain contact pads and e-beam lithography for gate channel definition. After cleaning the boron-doped (100) silicon samples with RCA solution, they are placed in an oxidation furnace to grow the gate oxide followed by N-type poly-silicon deposition. Since the devices are made on an individual basis, there is no need for extra steps for isolation as well as P and N-well definition. The oxide thickness is set to be 5nm for a channel length of 100nm. The sample is then coated with photoresist and by means of standard photolithography the gate contact pads are formed. Figure 5 displays, schematically, how the MOSFETs in this paper are being realized. Since we use positive resist, one step of lift-off is necessary to form the gate region. The lines corresponding to gate channel regions are drawn using the aforementioned nano-lithography technique. This step is essential for the formation of nano-scale MOSFET devices. After drawing lines, the photo-resist is developed in its standard developing solution and subsequent steps for the formation of MOSFET transistor are followed.

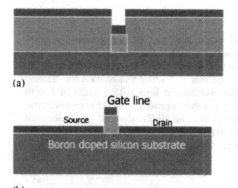

(a)

Gate line

Source   Drain

Boron doped silicon substrate

(b)

Figure 5: (a) The schematic showing the lift-off process for the formation of transistor gate region. (b) the evolution of source and drain area after the lift-off process has been completed and doping has also been performed by a Ge/Sb co-evaporation.

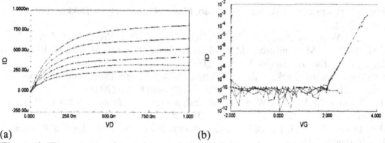

(a)   (b)

Figure 6: The current voltage characteristics of the transistors fabricated using this nano-lithography approach.

After the gate region is defined using our novel lithography approach, the gate oxide is patterned in dilute HF solution, followed by deposition of a thin layer of germanium-antimony layer to act as the solid diffusion source for the source-drain regions. The length of channel is defined using the width of the drawn lines with the lithography technique presented before and varies between 100 to 200nm for different samples. An inversion capacitance exceeding $0.7\mu F/cm^2$ was achieved for n-MOS transistors. The electrical characteristics of the device are collected in Figure 6. This figure shows the output and transfer characteristics of the transistors fabricated using this approach, indicating the efficacy of this lithography technique in making useful and functional transistors. The threshold voltage is measured to be -50mV (without threshold adjustment), drain-induced barrier lowering (DIBL) is measured to be 150mV/V and drive current equals to $310\mu A/\mu m$. These results suggest that the device has controllable short channel effect. Further characterization of the device is being pursued.

CONCLUSIONS

In summary, we have developed a novel lithography tool for applications in nano-electronic devices. This technique is essentially field emission of electrons from encapsulated carbon nano-tubes formed on silicon substrates. The encapsulation of the nanotubes leads to the formation of beam-shape emission and it can be used for drawing lines on a photo-resist coated substrate. The evolution of lines with a width of 100 to 200nm has been an indication for the usefulness of this approach. Also the lines have been used for the definition of channel gate of the MOSFET devices. The electrical characteristics of the MOSFET show a satisfactory behavior further corroborating the applicability of this approach in nano-electronic devices.

ACKNOWLEDGEMENT

This work has been supported by a grant from Iranian Ministry of Industry, and a grant from Natural Research Council of Canada.

REFERENCES

[1] Henk w. Ch. Postman, Tijs Teepen, Zhen Yao. Melena Grifoniu, Cees Dekker, Vol. 293 Science (2001).

[2] A. Javey, Q. Wang, A. Ural, Y. Li, H. Dai, Nano Letters (2002) Vol. 2, No.9, 929-932.

[3] J. Guo, S. Datta, M. Lundstrom, M.Brink, P. McEuen, A. Javey, H. Dai, H. Kim, P. McIntyre, IEEE Electron Device Meeting (2002).

[4] Jean- Marc Bonard, Mirko Croci, Christian Klinke, Fabine Conus, Imad Arfaoui, Thomas Stockli, and Andre Chatelaine; Physical Review(B) 67, (2003).

[5] Y. Li, D. Mann, M. Rolandi, W. Kim, Ant Ural, S. Hung, A. Javey, J. Cao, D. Wang, E. Yenilmez, , Nano Letters; Vol. 4, 317– 321 (2004).

[6] M.S. D resselhaus, G. Dresselhaus, ph. Avouris (Eds); Topics in Applied Physics, Spring, Verlag (2001).

[7] J. Guo, S. Datta, M. Lundstrom, M.Brink, P. McEuen, A. Javey, H. Dai, H. Kim, P. McIntyre; IEDM (2002).

[8] J. Koohsorkhi, H. Hoseinzadegan, S. Mohajerzadeh and M. Robertson, "Novel self-defined field-emission transistors with carbon nano-tubes on silicon substrates", IEEE DRC, USA, 2004.

[9] P. McEuen, M. Fuhrer, H. Park; IEEE Transactions on Nanotechnology, Vol. 1, March 2002.

[10] H. Hesamzadeh, B. Ganjipour, S. Mohajerzadeh, a. Khodadadi, Y. Mortazavi and S. Kiani, "PECVD-growth of carbon nanotubes using a modified tip-plate configuration", Carbon, vol. 42, no. 5-6, pp. 1043-1047, (2004)

[11] J. Koohsorkhi, Y. Abdi, S. Mohajerzadeh, H. Hoseinzadegan and A. Khakifirooz, "Self-defined vertically grown carbon nano-structures suitable for submicron and nano-lithography", Nano-tech, 2005

[12] Y. Abdi, S. Mohajerzadeh, J. Koohsorkhi and H. Hosseinzadegan, Applied Physics Letters, vol.88, 2006.

Mater. Res. Soc. Symp. Proc. Vol. 913 © 2006 Materials Research Society   0913-D04-08

Electron Thermal Transport Properties of a Quantum Dot

Xanthippi Zianni
Dept. of Applied Sciences, Technological Educational Institution of Chalkida, Psachna,
Chalkida, 34400, Greece

ABSTRACT

The electron thermal conductance of a dot has been calculated within a linear response theory in the regime of weak coupling with two electrode leads. Coulomb oscillations are found. We discuss the effect of the interplay between the charging energy, the thermal energy and the confinement in the behavior of the electron thermal conductance.

INTRODUCTION

For the past two decades structures based on semiconductor quantum dots have attracted a lot of research interest. Recently, there is an increased interest in studying the thermal properties in quantum dots structures [1-16]. In the recent years, there have been proposed thermoelectric applications of quantum dot superlattices made of different material systems as well as periodic arrays of dots[4,8-15]. Nanocrystalline silicon has recently proposed for designing efficient ultrasound emitter due to the measured low thermal conductivity relative to the bulk [5].

For a quantum dot the conductance for purely sequential tunneling has been investigated theoretically by Beenakker[17]. The developed theory has been extended to the thermopower by Beenakker and Staring [18] and it has been investigated experimentally by Staring et al [19]. The cotunneling regime [20] and the crossover have been studied by Turek and Matveev [21]. In the case of a quantum dot strongly coupled to one lead, the thermopower has been investigated by Matveev and Andreev [22]. Here, the electron thermal conductance of a quantum dot in the regime of weak coupling with two electrode leads is calculated within a linear response theory.

THEORY

We consider a double barrier tunnel junction. It consists of a quantum dot that is weakly coupled to two electron reservoirs via tunnel barriers. Each reservoir is assumed to be in thermal equilibrium and there are a voltage difference V and a temperature difference ΔT between the two reservoirs. A continuum of electron states is assumed in the reservoirs that are occupied according to the Fermi-Dirac distribution:

$$f(E - E_F) = \left[1 + \exp\left(\frac{E - E_F}{k_B T} \right) \right]^{-1}, \tag{1}$$

where the Fermi energy, E_F, in the reservoirs is measured relative to the local conduction band bottom.

The quantum dot is characterized by discrete energy levels E_p ($p=1,2,..$) that are measured from the bottom of the potential well. Each level can be occupied by either one or zero electrons. It is assumed that the energy spectrum does not change by the number of electrons in the dot. The energy levels are assumed to be weakly coupled to the states in the electrodes so that the charge of the quantum dot is well defined. We adopt the common assumption in the Coulomb blockade problems for the electrostatic energy $U(N)$ of the dot with charge $Q=-Ne$:

$$U(N) = (Ne)^2 / 2C - N\phi_{ext} \quad , \tag{2}$$

where C is the effective capacitance between the dot and the reservoirs and ϕ_{ext} is the external potential (e.g. the gate bias in the transistor configuration).

The tunneling rates through the left and right barriers from level p to the left and right reservoirs are denoted by Γ_p^l and Γ_p^r, respectively. It is assumed that energy relaxation rates for the electrons are fast enough with respect to the tunneling rates so that we can characterize the state of the dot by a set of occupation numbers, one for each energy level. It is also assumed that inelastic scattering takes place exclusively in the reservoirs not in the dot. The transport through the dot can be described by rate equations.

The energy conservation condition for tunneling implies the following conditions[17]:
- for tunneling from an initial state $E^{i,l(r)}$ in the left (right) reservoir to a final state p in the quantum dot:

$$E^{i,l}(N) = E_p + U(N+1) - U(N) + \eta eV \tag{3}$$

$$E^{i,r}(N) = E_p + U(N+1) - U(N) - (1-\eta)eV \tag{4}$$

- for tunneling from an initial state p in the quantum dot to a final state in the left (right) reservoir at energy $E^{f,l(r)}$:

$$E^{f,l}(N) = E_p + U(N) - U(N-1) + \eta eV \tag{5}$$

$$E^{f,r}(N) = E_p + U(N) - U(N-1) - (1-\eta)eV \tag{6}$$

where N is the number of electrons in the dot before the tunneling event, η is the fraction of the voltage V which drops over the left barrier. The energies in the reservoirs are measured from the local conduction-band bottom.

Due to the voltage difference V and the temperature difference ΔT between the two reservoirs, electric and thermal currents pass through the dot. The stationary current I and heat flux Q through the left barrier equals that through the right barrier and they are respectively given:

$$I = -e \sum_{p=1}^{\infty} \sum_{\{n_i\}} \Gamma_p^l P(\{n_i\}) \Big| \delta_{n_p,0} f(E^{i,l}(N) - E_F) - \delta_{n_p,1}[1 - f(E^{f,l}(N) - E_F)] \Big\} \quad (7)$$

$$Q = \sum_{p=1}^{\infty} \sum_{\{n_i\}} \Gamma_p^l P(\{n_i\}) \{ \delta_{n_p,0}[E^{i,l}(N) - E_F] f(E^{i,l}(N) - E_F) -$$
$$\delta_{n_p,1}[E^{f,l}(N) - E_F][1 - f(E^{f,l}(N) - E_F)] \} \quad (8)$$

where the second summation is over all possible combinations of occupation numbers $\{n_1, n_2, ...\} \equiv \{n_i\}$ of the energy levels in the quantum dot, each with stationary probability $P(\{n_i\})$. The numbers n_i can take on only the values 0 and 1. The non-equilibrium probability distribution P is a stationary solution of a kinetic equation. This has been solved in the linear regime by Beenakker [17]. The solution is substituted in equations (7) and (8) and the linearized expressions for the electric current I [17,18] and the heat flux Q are obtained:

$$I = \frac{e}{k_B T} \sum_{p=1}^{\infty} \sum_{N=1}^{\infty} \frac{\Gamma_p^l \Gamma_p^r}{\Gamma_p^l + \Gamma_p^r} P_{eq}(N) F_{eq}(E_p/N)[1 - f(\varepsilon_p - E_F)] \times$$
$$\left[eV - \frac{\Delta T}{T}(\varepsilon_p - E_F) \right] \quad , \quad (9)$$

$$Q = -\frac{1}{k_B T} \sum_{p=1}^{\infty} \sum_{N=1}^{\infty} \frac{\Gamma_p^l \Gamma_p^r}{\Gamma_p^l + \Gamma_p^r} P_{eq}(N) F_{eq}(E_p/N)[1 - f(\varepsilon_p)] \varepsilon_p \times$$
$$\left[eV + \frac{\Delta T}{T}(\varepsilon_p - E_F) \right] \quad , \quad (10)$$

where $\varepsilon_p \equiv Ep + U(N) - U(N-1) - E_F$. $P_{eq}(N)$ is the probability that the quantum dot contains N electrons in equilibrium and $F_{eq}(E_p/N)$ is the conditional probability in equilibrium that level p is occupied given that the quantum dot contains N electrons. The above equilibrium probabilities are respectively defined [17] as:

$$P_{eq}(N) = \sum_{\{n_i\}} P_{eq}(\{n_i\}) \delta_{N, \sum_i n_i} , \quad (11)$$

$$F_{eq}(E_p/N) = \frac{1}{P_{eq}(\{n_i\})} \sum_{\{n_i\}} P_{eq}(\{n_i\}) \delta_{n_p,1} \delta_{N, \sum_i n_i} . \quad (12)$$

$P_{eq}(\{n_i\})$ is the Gibbs distribution in the grand canonical ensemble:

$$P_{eq}(\{n_i\}) = Z^{-1} \exp\left[-\frac{1}{k_B T}\left(\sum_{i=1}^{\infty} E_i n_i + U(N) - NE_F\right)\right],$$ (13)

where $N \equiv \sum_i n_i$, and Z is the partition function:

$$Z = \sum_{\{n_i\}} \exp\left[-\frac{1}{k_B T}\left(\sum_{i=1}^{\infty} E_i n_i + U(N) - NE_F\right)\right].$$ (14)

The thermal conductance is defined as:

$$\kappa \equiv -\frac{Q}{\Delta T}\bigg|_{I=0} = -K\left(1 + \frac{S^2 GT}{K}\right).$$ (15)

where the conductance G, the thermopower S and the thermal coefficient K are respectively defined as:

$$G \equiv \frac{I}{V}\bigg|_{\Delta T=0},$$ (16a)

$$S \equiv -\frac{V}{\Delta T}\bigg|_{I=0},$$ (16b)

and

$$K \equiv \frac{Q}{\Delta T}\bigg|_{V=0}.$$ (16c)

RESULTS

Coulomb effects are important when the charging energy dominates over the thermal energy and the electron confinement, i.e. $e^2/C > k_B T, \Delta E$. We discuss the behavior of the electron thermal conductance in three regions: in the quantum limit ($\Delta E >> k_B T$), in the classical region ($\Delta E << k_B T$) and in the intermediate energy region.

In the quantum limit, where $\Delta E >> k_B T$, the discreteness of the energy spectrum of the quantum dot plays a predominant role. The calculated thermal conductance is plotted in figure 1 for two values of the ratio $\Delta E/k_B T$ in the quantum limit. For the sake of simplicity, it has been assumed an equidistant energy levels spectrum ($E_p = p\Delta E$) and level-independent tunneling rates, i.e. $\Gamma_p^{l,r} = \Gamma^{l,r}$ .the thermal conductance exhibits periodic Coulomb-blockade oscillations. The peaks occur each time an extra electron enters in the dot and they are separated by intervals $\Delta E_F = \Delta E + \dfrac{e^2}{C}$. The same periodicity has been found elsewhere [17,18] for the conductance, G, and the thermopower, S, of quantum dots.

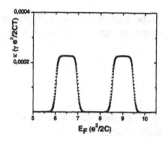
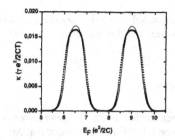

Figure 1. Calculated electron thermal conductance, κ, for a series of equidistant, nondegenerate levels with separation $\Delta E = 0.5\ e^2/2C$ and for $k_BT = 0.05\ e^2/2C$ (left figure) and $k_BT = 0.1$ $e^2/2C$ (right figure).

The shape of the peaks and the magnitude of the maximum exhibit a dependence on the ratio $\Delta E/k_BT$. The electron thermal conductance of a quantum dot decreases rapidly with decreasing temperature and increasing ΔE.

In the classical regime, $\Delta E \ll k_BT$, the discreteness of the energy spectrum of the quantum dot is screened by the thermal energy and the energy spectrum can be treated as a continuum. The calculated thermal conductance in the classical regime is shown in figure 2 for different temperatures. Figure 2a shows the behavior of κ at low temperatures (big charging energies). The periodicity of κ is the same as that of the conductance, G. The effect of increasing thermal energy is shown in figure 2b. The electron thermal conductance is non zero between the peaks, due to thermal broadening of the distribution functions. The peaks of κ are located at the valleys of G.

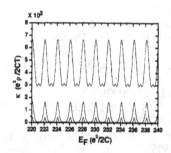
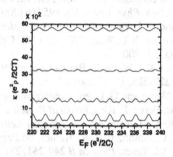

Figure 2. Coulomb oscillations of the electron thermal conductance in the classical regime as a function of temperature: (a) for $k_BT = 0.05, 0.1, 0.2\ e^2/2C$ (left figure) and (b) $k_BT = 0.1, 0.2, 0.3,$ $0.4, 0.5\ e^2/2C$ (right figure). The order of the curves corresponds to increasing temperature. The lowest curve corresponds to the lowest temperature.

The intermediate region extends between the quantum regime and the classical region. Here, the thermal energy and the levels spacing are comparable. The thermal energy is lower than the charging energy ($k_BT < e^2/C$) and Coulomb effects are non-negligible. Similar qualitative behavior has been found as that shown in figure 2. However, the electron thermal conductance differs quantitatively from the classical one because the distributions are different in the two regions. Morevover, the discreteness of the energy spectrum cannot neglected in the intermediate region..

ACKNOWLEDGMENTS

The present work has been funded by the European Commission and by the Greek Ministry of Education, O.P. 'Education' (E.P.E.A.E.K.), under the 'Archimides' programme.

REFERENCES

1. T.C. Harman, M.P. Walsh, B.E. Laforge, G.W. Turner, *Journal of Electronic Materials* **34**, L19 (2005).
2. A. Baladin, *Journal of Nanoscience and Nanotechnology* **5**, 1015 (2005).
3. V. Sajfert, J. P. Setrajcic, S. Jacimovski, B. Tosic, *Physica E* **25**, 479 (2005).
4. Y. Bao, W.L. Liu, M. Shamsa, K. Alim, A.A. Balandin, J.L. Liu, *Journal of the Electrochemical society* **152**, G432 (2005).
5. T. Kihara, T. Harada, N. Koshida, *Japanese Journal of Applied Physics Part 1* **44**, 4084 (2005).
6. M. Shamsa, W. Liu, A.A. Balandin, J. Liu, *Applied Physics Letters* **87**, 202105 (2005).
7. M.C. Llaguno, J.E. Fischer, A.T. Johnson, J. Hone, *Nano Letters* **4**, 45 (2004).
8. D. Vashaee and A. Shakouri, *Phys. Rev. Lett.* **92**, 106103 (2004).
9. R. Yang and G. Chen, *Phys. Rev. B* **69**, 195316 (2004).
10. A.A. Balandin and O. Lazarenkova, *Applied Physics Letters* **82**, 415, (2003).
11. J.L.Liu, A.Khitun and K.L. Wang, W.L. Liu, G. Chen, Q.H. Xie, S.G. Thomas, *Physical Review B* **67**, 165333 (2003).
12. J.L.Liu, A. Khitun, K.L.Wang, T. Borca-Tasciuc, W.L. Liu, G. Chen, D.P. Yu, *Journal of Crystal Growth* **227-228**, 1111 (2001).
13. A. Khitun, K.L. Wang and G. Chen, *Nanotechnology* **11**, 327 (2000).
14. J. P. Small, K. M. Perez, P. Kim, *Physical Review Letters* **91**, 256801 (2003).
15. A.V. Andreev, K.A. Matveev, *Physical Review Letters* **86**, 280 (2001).
16. A.S. Dzurak, C.G. Smith, C.H.W. Barnes, M.Pepper, L. Martin-Moreno, C.T. Liang, D.A. Ritchie, G.A.C Jones, *Physica B* **249-251**, 281 (1998).
17. C.W.J. Beenakker, *Phys. Rev. B* **44**, 1646 (1991).
18. C.W.J. Beenakker and A.A.M. Staring, *Phys. Rev. B* **46**, 9667 (1992).
19. A.A.M. Staring, L.W. Molenkamp, B.W. Alpenaar, H. van Houten, O.J.A. Buyk, M.A.A. Mabesoone, C. W.J. Beenakker and C.T. Foxon, *Europhys. Lett.* **22**,57(1993).
20. D.V. Averin and Y.V. Nazarov, *Phys. Rev. Lett.* **65**, 2446 (1990).
21. M.Turek and K.A. Matveev, *Phys.Rev. B* **65**, 115332 (2002).
22. K.A. Matveev and A.V. Andreev, *Phys. Rev B* **66**, 045301 (2002).

Modeling and Metrology

Mater. Res. Soc. Symp. Proc. Vol. 913 © 2006 Materials Research Society   0913-D05-01

Using Quantitative TEM Analysis of Implant Damage to Study Surface Recombination Velocity in Silicon

Jennifer Lee Gasky, Sophya Morghem, and Kevin Jones
Materials Science and Engineering, University of Florida, 100 Rhines Hall, P.O. Box 116400, Gainesville, FL, 32611-6400

ABSTRACT

Silicon wafers with shallow trench isolation structures 3700Å deep were self-implanted with silicon at 40keV, and a dose of 1E15/cm^2. This produced an amorphous layer 1000Å deep. The samples were subsequently annealed at temperatures ranging from 750°C to 900°C. The excess interstitials can recombine at the "surface" created by the proximity to the trench sidewall. Plan-view TEM was used to quantify the dislocation distribution as a function of distance from the trench sidewall. It was found that there was no measurable change in defect density as a function of distance from the trench. This was true for both the 20 minute isochronal anneal, and the isothermal study 750°C. This suggests there is a relatively weak recombination of interstitials at the surface. This is surprising given most of the TCAD models assume a very fast surface recombination velocity.

INTRODUCTION

In order to model advanced microelectronic devices, an accurate understanding of interstitial recombination at the surface in necessary. The goal of the experiment was to use the evolution of end of range damage in patterned silicon to develop a better understanding of the strength of surface recombination. Ion implantation in semiconductor devices is known to introduce point defects. Upon annealing, these point defects evolve into {311} defects and subsequently dislocation loops. During this process the interstitial population decreases. Surface recombination is believed to account for much of the loss of these interstitials.

Previously, surface recombination velocities were measured through various techniques, including the measurement of diffusion near the surface and the monitoring of the dissolution of defects. Previous results on the role of the surface are conflicting. Papers by B. Colombeau et al suggest the surface acts as a strong sink for interstitials. This paper shows a lack of diffusion in the first layer of a boron superlattice. Slow diffusion was attributed to the strong surface recombination of interstitials.[1]

However both Ganin and Marwick[2] and King et al[3] showed that lapping the surface to reduce the amorphous layer thickness of an amorphizing implant(there-by bringing the end of range(E.O.R.) damage closer to the surface) had no effect on the evolution of E.O.R. damage upon annealing. King's experiment involved four samples chemically and mechanically lapped to depths of 180Å, 155Å, 125Å, and 80Å. The results showed no difference in defect evolution, which suggests the surface does not act as a strong sink.

However, in both cases an amorphous layer was present between the E.O.R. and the surface until re-crystallization was complete. This experiment attempts to study surface recombination laterally at the trench sidewall since no amorphous layer exists between this "surface" and the E.O.R. damage.

EXPERIMENTAL

The wafers were self-implanted at 40keV and a dose of 1E15/cm^2. Trenches were patterned into a Si wafer 4400Å deep, and self-implanted to create an amorphous layer 1000Å thick. Cross-section TEM was used to verify the amorphous layer thickness and the presence of regrowth related defects along the sides(Figure 1). These defects along the sidewall form because the amorphous Si cannot re-crystallize defect free at the triple phase point of the oxide, amorphous Si and crystalline Si. The defects from solid-phase-epitaxial-regrowth(SPER) along the sidewall made it impossible to count the defects within ~150Å of the trench.

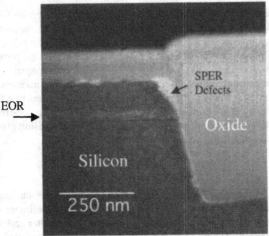

Figure 1: XTEM image of a sample annealed at 700°C for one minute.

Plan-view TEM was used in the analysis portion of the project to quantify the defects as a function of lateral distance from the trench. An example of a typical square pattern that was used to quantify defect density can be seen in Figure 2. To obtain statistical data and determine dependence, two studies were conducted: isothermal and isochronal.

Trench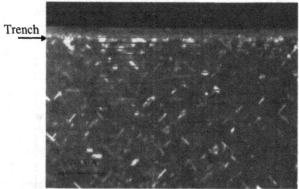

Figure 2: Plan-view TEM image displaying surface, depth and defects.

The isothermal study was conducted at 750°C for annealing times ranging from 20 minutes to 240 minutes. For the isochronal study, the temperatures ranging from 750°C to 950°C at 50°C increments were used to anneal the samples for 20 minutes.

Sample preparation included coring, annealing, etching, and finally TEM analysis. The sample was cored from the dye-pattern on the wafer. Standard backside etching was used to create the TEM samples. TEM analysis was done using weak beam dark field g_{220} imaging conditions. Once negatives were obtained from the TEM camera, prints were developed for a total magnitude of 136,206.9 X. In order to assess and quantify the defect density for each sample, a counting system was created using a rectangular grid. This rectangular grid maintained a width of 147Å per rectangle for a total of 14 rectangles per grid. So the total lateral distance into each sample was 2055Å. Two different lengths of the grids were chosen for each of the studies. For the isochronal study, the length of the grid was maintained at 3670Å, while the isothermal study the length of the grid was 6610Å. For the counting procedure, the first rectangle of the grid was ignored due to the SPER defects. For a given sample, the defects present in each rectangle were measured in size and multiplied by the density of interstitials in each type of defect counted. If the loop was circular or elliptical in nature, the diameter was used, however if the loop was rectangular in nature, the width and length of the loop was recorded. Each rectangle was counted three times for a statistical average. The results were plotted as the interstitial concentration in each rectangle versus the distance from the trench edge.

RESULTS AND DISCUSSION

Figure 3 shows the results from the sample annealed at 750°C for 20 minutes. Trapped interstitial population shows no evidence of decreasing down to a distance of 170Å from the trench sidewall.

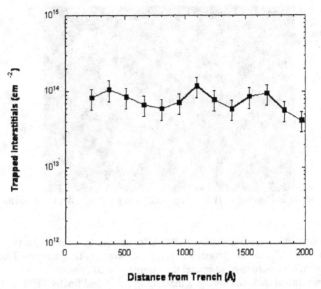

Figure 3: Sample annealed at 750°C for 20 minutes.

Figure 4 shows some of the images of the defect evolution for the isothermal annealing study. The (311)'s are going through their complete evolution to loops and there is no obvious signs of enhanced defect dissolution closer to the trench.

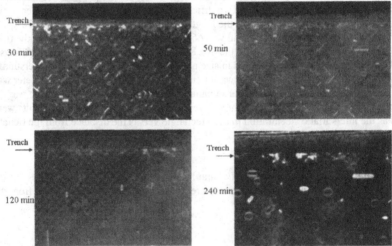

Figure 4: Isothermal Study: Samples annealed at 750°C for 30, 50, 120, and 240 minutes.

A graph of the entire isothermal anneal is shown in Figure 5. These plots show no strong evidence of surface recombination. As mentioned earlier, the results of Colombeau suggest surface recombination is significant for depths up to 2500 Å. If there was strong surface recombination, a drop in interstitial density would be expected as the distance gets closer to the edge. But as seen in the graph, the density does not drop off for any of the anneal times.

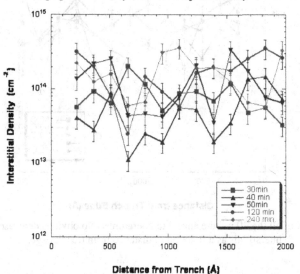

Figure 5: Plot of all the 750°C anneal times. No obvious decrease in trapped interstitials as a function of distance from trench edge.

Figure 6 shows the results of the 20 minute isochronal annealing study. Again the defect density does not vary with distance from the trench edge, suggesting recombination at this surface is weak.

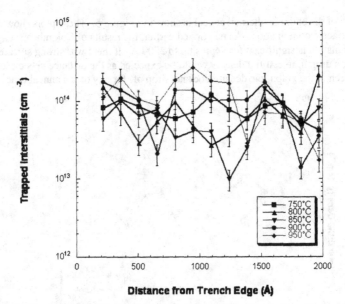

Figure 6: Plot of all the 20 minute anneal temperatures. No obvious decrease in trapped interstitials as a function of distance from trench edge.

Conclusions

Using Si wafers with STI structures and with 40keV amorphizing implants, isothermal and isochronal studies were conducted to investigate surface recombination of interstitials as a function of lateral distance from a trench edge by quantifying the trapped interstitial density in extended defects. There was no decrease in trapped interstitial density as a function of distance from the lateral sidewall surface for distances between 180Å and 2000Å. These studies suggest lateral recombination at the side of the trench does not have a measurable effect on the trapped interstitials for distances >180Å. These results also suggest that surface recombination at the sidewall is not strong enough to effect defect evolution.

References

1 B. Colombeau et al. Mat. Res. Soc. Symp. Proc. 810, 91(2004)
2 E. Ganin and A. Marwick, Mater. Res. Soc. 147,13 (1989)
3 A.C. King et al. *Journal of Applied Physics*, **93**, 244 (2003)
4 Alain Claverie et al. Mat. Res. Soc. Symp. Vol. 610 (2000)
5 M.E. Law et al. Journal of Applied Physics, 84, 7 (1998)
6 D.R. Lim, C.S. Rafferty, and F. P. Klemens, Appl. Phys. Lett. 67, 2302 (1995)

Mater. Res. Soc. Symp. Proc. Vol. 913 © 2006 Materials Research Society   0913-D05-02

Diffraction from Periodic Arrays of Oxide-Filled Trenches in Silicon: Investigation of Local Strains

Michel Eberlein[1,2], Stephanie Escoubas[1], Marc Gailhanou[1], Olivier Thomas[1], Pascal Rohr[2], and Romain Coppard[2]

[1]Laboratoire TECSEN UMR 6122 CNRS Case 262, Université Paul Cézanne, 54 Avenue Normandie-Niemen, 13397 MARSEILLE Cedex 20, France

[2]ATMEL ROUSSET, Zone Industrielle, 13106 ROUSSET Cedex, France

ABSTRACT

The experimental evaluation of stresses at the nanometer scale is a real challenge. We propose an innovative use of High Resolution X-Ray Diffraction to measure local strains induced in silicon by periodic arrays. This technique is non-destructive and allows for the measurement of periodic strain fields in monocrystalline silicon, created in particular by Shallow Trench Isolation process. A 0.58μm-period array of trenches filled with SiO_2 gives rise to satellites in reciprocal space maps around the Si substrate peak. The intensity and envelope of these satellites depend on the local strain. The experimental reciprocal space maps are compared to those computed using the kinematical theory from the elastic displacement field calculated with finite element modelling. This technique allows us to study the generation of strain during the main steps of the STI process. During the process, after trenches get filled with oxide and top layers removed, a second diffraction peak appears with a lower intensity than the substrate one. Thanks to finite element modelling, we validate that this peak is caused by an almost constant strain in the silicon active areas. Typical values of strains after trench filling are $\varepsilon xx = -1.68*10^{-3}$ and $\varepsilon zz = 1.56*10^{-3}$ where x and z refer to transverse and perpendicular directions.

INTRODUCTION

With the constant decrease of critical size in microelectronics, mechanical stresses in thin films and nanostructures have become an important matter. They may be used as a benefit, for example by increasing the electron mobility [1] or on the opposite have real hazardous effects like dislocation generation or leakage current increase [2]. In order to improve integration and specification of components, Shallow Trench Isolation (STI) technology is used in many microelectronics applications like non-volatile memories [3]. Deep trenches are etched in the silicon substrate and filled with SiO_2 in order to isolate electrically the memory cells. The many steps of the process generate high mechanical stresses, which may damage the device and reduce its reliability, more particularly with the decrease in critical dimensions. This is the reason why it is crucial to measure stress at the local scale. While many techniques can yield average stress, it is a real challenge to get local strain fields with a nanometer spatial resolution. For instance, microdiffraction or micro-Raman spectroscopy are limited to a spatial resolution of 0.3μm [4,5]. Convergent Beam Electron Diffraction has the nanometer resolution but needs a painful preparation that can modify the strain state in the sample [6]. High Resolution X-Ray Diffraction (HR-XRD) combines the advantages of being very sensitive to local strain field and also non-destructive. In this study we have used HR-XRD from periodic arrays [7,8]. The Fourier pattern resulting from the periodicity is obtained by summing the amplitude from each single cell. In this way one can extract local information from individual patterns. Elastic displacement fields calculated with Finite Element Modelling (FEM) are used to calculate Reciprocal Space Maps (RSMs) which are compared to the experimental ones [9].

EXPERIMENT

A 4-reflection Ge 220 DuMond-Bartels monochromator and a 3-reflection Ge 220 analyser were used on a 4 circle-goniometer with a sealed Cu X-Ray tube. With this setup both monochromator and analyser streaks are reduced and almost not visible in the RSMs. The incoming beam has a wavelength $\lambda_{CuK\alpha 1}$= 0.1540598 nm and a divergence in the scattering plan of 12 arcsec after the DuMond-Bartels monochromator. For this study, high resolution 004 symmetric and 404 asymmetric RSMs were measured. The three components (along the three crystallographic axes of silicon) of the scattering vector using the reduced components H, K and L write

$$q_x = H * \frac{2\pi}{a} \qquad\qquad q_y = K * \frac{2\pi}{a} \qquad\qquad q_z = L * \frac{2\pi}{a}$$

where a is the silicon lattice parameter (a=0.543088nm).

[100] oriented trench arrays etched in silicon (001) were analysed. The trenches are 410 nm deep and filled with SiO_2 (Figure 1a). The samples ($2*2mm^2$) have a periodicity of 0.58μm with a Si active area of 0.22μm. Scanning Electron Microscopy cross section (Figure 1b) show that trench walls are not vertical. This is done intentionally in order to avoid stress singularities in silicon, which may induce nucleation of dislocations.

In order to study the generation of strains during STI fabrication, four samples taken out after key process steps were analysed. The first step in the process is a silicon etching (Figure 2, sample 1). There are already deposited layers on top of the active area but trenches are still empty. The second step is a thermal oxidation which creates a 15nm thick oxide liner (sample 2). Then, the trenches are filled with deposited SiO_2 and Chemical and Mechanical Polishing (CMP) of the top layer (sample 3) is performed. Finally, the oxide is densified at 850°C.

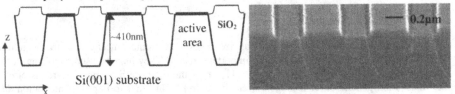

Figure 1: Schematic representation (a) and SEM cross-section (b) of isolation trenches

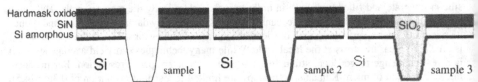

Figure 2: schematic representation of STI trenches: samples 1) after etching, sample 2) after liner oxide, sample 3) after trenches filling and CMP

MODELLING

For this study, all calculations were performed with a FEM software (FEMlab®) in plane strain conditions (along x and z directions). The structures are considered as infinite in the y direction compared to the period. Hypothesis for calculations of the strain for one period are the following:

- All materials are assumed to be elastic and isotropic.
- Strain is applied in materials only through thermal cooling with
  $\varepsilon = [\alpha(Si) - \alpha(SiO_2)] * \Delta T$
- Lateral borders are blocked along x direction

The last condition is a simulation convenience to introduce periodicity in the calculation even if we only consider one period of the pattern. The strain values obtained with FEM at any point of the structure are then used to calculate a theoretical diffraction spectrum which will be compared to fit with experimental results.

The diffracted intensity, in kinematic approximation, is proportional to:

$$I(\vec{q}) \propto |F(\vec{q})|^2 = \left| \sum_j f_j(\vec{q}).e^{i\vec{q}.(\vec{r}_j + \vec{u}_j)} \right|^2$$

where $\vec{r}_j$ et $\vec{u}_j$ are respectively the position and displacement of atom j inside the full structure, $\vec{q}$ is the scattering vector and f_j the atomic scattering factor. In the particular case of a sample periodic in the x direction, the diffracted intensity is proportional to the squared modulus of the structure factor of a single period $F_0(\vec{q})$ multiplied by an expression representative of the periodicity:

$$|F(\vec{q})|^2 = |F_0(\vec{q})|^2 \left[\frac{\sin^2(\frac{Nq_x\Lambda}{2})}{\sin^2(\frac{q_x\Lambda}{2})} \right]$$

where $N\Lambda$ is the coherence length of the beam. It is important to note that the information on the local strain field lies in the envelope function $|F_0(\vec{q})|^2$

RESULTS AND DISCUSSION

Asymmetric 404 reflection scans were carried out on all the samples to create RSMs of the plane perpendicular to trenches. Just after silicon etching, sample 1 mainly shows an intense Bragg diffraction peak and satellite rods along the L direction (Figure 3). The Bragg peak is attributed to the diffraction of the unstrained silicon substrate. The satellites' spacing is inversely proportional to STI period. The 0.58μm period extracted from the map is in perfect agreement with the one measured on SEM cross section. The satellites' envelope is almost centred on the substrate diffraction peak. This indicates that the strain is small but yet visible as there is a small dissymmetry in the RSM. To be sure that this is not a measurement artifact, all the top layers have been removed by HF dip before the reciprocal space measurement (Figure 4).

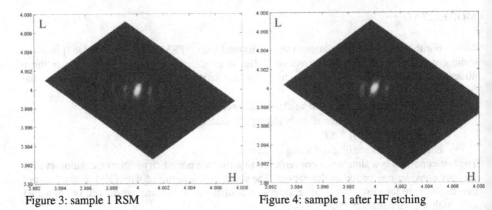

Figure 3: sample 1 RSM   Figure 4: sample 1 after HF etching

Satellite rods are well aligned along H direction and centred on the Bragg peak as the last one but in this case the map is perfectly symmetric. The slight dissymmetry observed on sample 1 RSM can thus be attributed to the top layers. Figure 5 shows the experimental 404 Si RSM after the 15nm thermal oxidation (sample 2). The asymmetry and the shift in satellites' intensity is directly related to the strain state of silicon. The sign of the strain is indicated by the intensity maximum position compared to the Si substrate peak. In that case the satellites' intensity maximum appears at larger H value, which is due to compressive strain across the trenches ($\varepsilon_{xx}<0$). The lower L value is related to tensile strain in the vertical direction ($\varepsilon_{zz}>0$). Even if the effect of the strain is already visible on X-Ray maps, the value is still quite low ($<5*10^{-4}$). After SiO_2 trench filling (sample 3), the satellites rods appear on the RSM on a wider range along x and z directions. One can distinguish a 'zone' where satellites' intensity is enhanced. The shift is in the same direction that in the previous sample ($\varepsilon_{xx}<0$ and $\varepsilon_{zz}>0$) but is more pronounced (Figure 6), indicating higher strain. After oxide densification at 850°C, the result is similar to sample 3, which indicates that this step does not modify significantly the strain state in silicon.

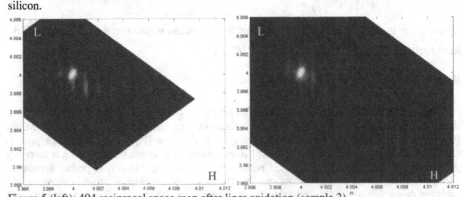

Figure 5 (left): 404 reciprocal space map after liner oxidation (sample 2)
Figure 6 (right): 404 reciprocal space map after trench filling and CMP (sample 3)

One can conclude that most of the strain is introduced in silicon by the High Density Plasma (HDP) oxide used for trench filling. All these results are confirmed by simulations (not shown here). 404 RSMs were also performed with the scattering plane along the lines. There is no visible effect of any strain in the y direction. This confirms the plane strain hypothesis used for FEM. Further in the process, SiN and amorphous Si top layers are removed. The experimental 404 RSM (Figure 7a) shows a particularity: a secondary diffraction peak. The same feature is found on the simulated map (Figure 7b), which shows good agreement with the experimental one. The maximum position is found at H= 4.0067 and L= 3.9938, which yields ε_{xx}= -1.68*10^{-3} and ε_{zz}= 1.56*10^{-3}. This suggests there is a large enough amount of silicon with almost constant strain to produce a well separated diffraction peak in the grating satellites' envelope. To confirm this hypothesis we studied the strain field simulated with FEM (Figure 8a). The variation of ε_{xx} and ε_{zz} values with depth, extracted in the middle of the silicon active area, is plotted on figure 9b. It is observed that strain values corresponding to the secondary peak match with plateaus or slowly varying of strains in silicon. ε_{xx} and ε_{zz} remains stable at +/- 10^{-4} on a depth of 180nm, then it decreases with increasing depth to become quite low in the substrate (ε<5.10^{-4} for depth >480nm). This confirms that there is a silicon area with homogeneous strain.

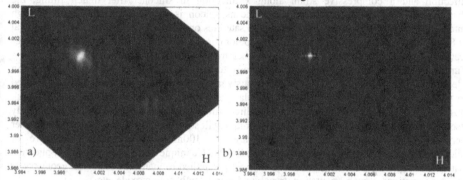

Figure 7: 404 RSM after trench filling and top layers removal a) experimental b) simulated

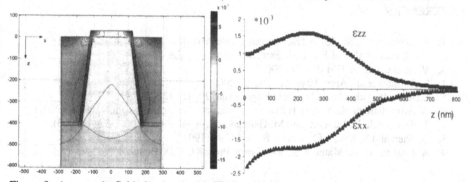

Figure 8: a) ε_{xx} strain field simulated with FEM, solid lines are isostrain curves and b) variation of ε_{xx} and ε_{zz} with depth in the middle of silicon active area extracted from a)

Even if strains can be higher near singularity points (trenches corners, etc.), this value is quite representative of the global strain in the silicon active area. It is a very interesting result since one can directly extract a strain value from the experimental results without complicated simulation.

As mentioned before, the HDP oxide used for the trench filling is responsible for most of the strain in that case and the results obtained in this study are in agreement with previous analysis done on oxide filling [1]. Also, the properties of the filling oxide could greatly influence the strain generation process. For instance, STI Gap-Fill Technology with High Aspect Ratio Process (HARP) has proved to change drastically the strain field [1] in silicon active areas.

CONCLUSIONS

X-Ray Diffraction is used to investigate the strain field generated during the main steps of STI production. High resolution RSMs provide experimental measurements of the strain at the nanometer scale. The maps show satellites along the H direction due to the periodicity. After liner oxidation, the dissymmetry in the satellites intensity is attributed to a tensile strain along z direction and a compressive strain along x direction in silicon. After trench filling, the dissymmetry is even more visible, which means silicon is in a higher strain state. A high intensity 'zone' can clearly be distinguished in satellites on RSMs. After the densification anneal at 850°C, the reciprocal space map shows no major change. Even if small strains are visible after trench oxidation, they are mainly introduced when trenches are filled with oxide.

Further in the process, the strain field gives raise to a second diffraction peak, which has been identified as the diffraction from a homogeneously strained area in active silicon. FEM was used to simulate the displacement field and calculate reciprocal space maps. The secondary peak position is confirmed by the simulation. This direct strain value measurement will be particularly useful to compare different samples. This technique will soon be applied to samples with smaller period (0.2μm) and with silicon active area of 100nm and 50nm, representative of microelectronics new generations devices. Finally, this technique has a wide range of application with a good spatial and strain sensitivity. It could be applied to many samples with geometry or process variations: trench orientation, trench width and depth, filling oxide, etc.

REFERENCES

1. R. Arghavani et al., IEEE Transactions on electron devices 51, 1740 (2004).
2. M. H. Park et al., IEEE IEDM, Extended Abstracts, 669 (1997).
3. V. Senez et al., IEDM 01-831 (2001).
4. N. Tamura et al., Appl. Phys. Lett. 80, 1 (2002).
5. I. De Wolf et al., Microelectronic Engineering 70, 425 (2003).
6. A. Armigliato et al., Materials Science in Semiconductor Processing 4, 97-99 (2001).
7. T. Baumbach, D. Lübbert, and M. Gailhanou, J. Appl. Phys. 87 (8), 3744 (2000).
8. Q. Shen and S. Kycia, Phys. Rev. B 55, 15791 (1997).
9. A. Loubens et al., Mater. Res. Soc. Symp. Proc. 875 (2003).

Mater. Res. Soc. Symp. Proc. Vol. 913 © 2006 Materials Research Society   0913-D05-03

Stress and Strain Measurements in Semiconductor Device Channel Areas by Convergent Beam Electron Diffraction

Jinghong Li, Anthony Domenicucci, Dureseti Chidambarrao, Brian Greene, Nivo Rovdedo, Judson Holt, Drerren Dunn, and Hung Ng
Micro-electronics_STG, IBM, 2070 Route 52, Mail Stop 40E, Hopewell Junction, NY, 12533-6531

ABSTRACT

Convergent electron beam diffraction (CBED) has been successfully applied to measure strain/stress in the channel area in PMOS semiconductor device with embedded SiGe (eSiGe) for 65nm technology. Reliable results of strain/stress measurements in the channel area have been achieved by good fitting of experimental CBED patterns with theoretical calculations. Stress measurements from CBED are in good agreement with simulations. A compressive stress as high as 823.9 MPa was measured in the <110> direction in the channel area of a PMOS device with eSiGe with 15% Ge and a thickness of 80nm. Stress measurements from CBED also confirm that the depth of the eSiGe and defects such as dislocation loops within the eSiGe relax strain/stress within the film and reduce strain/stress in the channel area.

INTRODUCTION

Traditional CMOS scaling which is mainly based on shrinking the geometric dimensions of devices is nearing to its end. This has led the semiconductor industry to turn to strain based engineering and new materials innovations as a means to enhance device performance. It is well known that PMOS performance can be improved by applying compressive stress to the channel while NMOS performance is improved by tensile stress. In particular, embedded SiGe grown in the recessed source/drain regions has been successfully used as a stressor to enhance PMOS hole mobility and device performance.[1-4] With the emphasis on stress in device channel areas, the demand for strain/stress measurements with high special resolution has increased. The ability to measure strain/stress in the channel areas of CMOS devices can provide better understanding of device performance enhancement. CBED has proved to be a powerful tool to measure local strain on nano-scale due to its high sensitivity and high spatial resolution. The local crystal lattice parameters (and hence strain) can be determined from CBED patterns by measuring shifts in high order Laue zone (HOLZ) lines, exposed to an electron beam of a few nanometers. Eventually, strain in crystal could be measured.[5] Several researchers has used CBED to measure strain/stress in semiconductor device structures, such as in Si near shallow trench isolation, source/drain regions and in channel area caused by NiSi and CoSi contacts.[6-8] This paper describes the application of CBED in measuring strain/stress in channel area in semiconductor device where eSiGe has been incorporated into source/drain (S/D) regions as a stressor. The influences of defects at the eSiGe/Si interface or within the eSiGe and depth of recess eSiGe on stress in channel area will be discussed. Experimentally determined stresses will also be compared with the device modeling results.

EXPERIMENTAL

The samples studied in this work were pFET devices fabricated with embedded SiGe in the source/drain regions to stress the channel region. After gate formation, a selective recess etch process was used to etch the source/drain regions of PFETs. Intrinsic $Si_{0.85}Ge_{0.15}$ was selectively grown in the recessed S/D regions. Then B implant was employed either as PFET S/D implant or as contact implant. Finally, NiSi was formed over eSi_xGe_{1-x} regions. Cross-sectional transmission electron microscope (XTEM) samples were prepared either by polishing followed by normal ion beam milling or by focused ion beam (FIB) to a thickness between ~250nm and ~350nm. Thickness of cross-section TEM sample was measured either by a projection method under two beam conditions or by electron energy loss spectrometry (EELS). CBED analysis was performed in a JEOL 2010F TEM equipped with a Gatan energy-filter, in the scanning transmission (STEM) mode with a probe size of 1 nm probe. Energy filtering was used by to improve sharpness of HOLZ lines by eliminating the effects of inelastically scattered electrons. The CBED experiments were performed by tilting the TEM samples along the <400> band to either the <230> zone axis (11.3°) or the <430> zone axis (8.3°). CBED patterns were recorded from 25nm to 1000nm below the gate at various intervals. The patterns were then analyzed using the ASAC software developed by SIS.[9-10], using the CBED pattern at 1000nm below the gate as an unstrained Si reference. The strain tensor in the Si crystal reference system was determined using the Chi^2 values determined by the software to ensure reliable calculation results.[5, 11] The stress in Si crystal system was calculated by Hook's law. The stress was transformed into a reference system, **theta$_{sample}$**, where the current flow direction of the device was one of the major axes, using standard tensor analysis as the following:

$$\textbf{theta}_{sample} = S_\sigma \textbf{theta}_{cryst}$$

where S_σ is the stress tensor rotation matrix and **theta$_{cryst}$** is the stress in the Si crystal system.

RESULTS AND DISCUSSION

Figure 1 shows a bright field XTEM image of PFET gate with eSiGe at the S/D region. No crystallographic defects were seen in the eSiGe or at the eSiGe/Si interface, indicating that the epitaxial crystal quality of eSiGe was very good and highly strained.

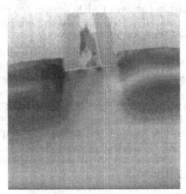

Figure 1 Bright field XTEM image of PFET gate with eSiGe at the S/D region.

Figure 2 shows (a) a dark field STEM image of PFET gate and four typical CBED patterns taken on the <230>zone axis at different depths below the gate. No obvious HOLZ line splitting was observed in the CBED patterns, indicating no surface relaxation due to thinning of TEM sample. The results from the ASAC analysis are shown in Table1:

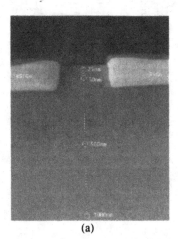

(a)

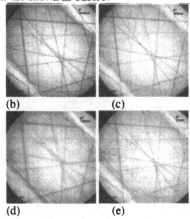

(b)   (c)

(d)   (e)

Figure 2 (a) Dark field STEM image of eSiGe gate and four CDED patterns at different depths: (b) 1000nm; (c) 500nm; (d) 100nm and (e) 25nm.

Table1

Position (nm)	Chi^2	V_{eff} (keV)	a	b	c	alpha	beta	gamma
1000	1.40	199.21	5.4310	5.4310	5.4310	90.00	90.00	90.00
500	0.70		5.4300	5.4300	5.4314	90.00	90.00	89.98
300	0.76		5.4300	5.4300	5.4334	90.00	90.00	89.98
200	1.10		5.4290	5.4290	5.4324	90.00	90.00	89.96
150	1.10		5.4291	5.4291	5.4313	90.01	89.99	89.96
100	4.20		5.4280	5.4280	5.4324	89.99	90.01	89.92
75	2.50		5.4242	5.4242	5.4354	90.02	89.98	89.96
50	1.30		5.4222	5.4222	5.4384	90.00	90.00	89.80
25	3.50		5.4180	5.4180	5.4404	90.02	89.98	89.72

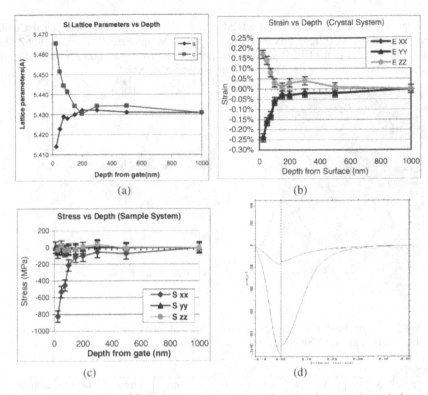

(a) (b)

(c) (d)

Figure 3 (a) variations of Si lattice parameters vs. depth below the gate, (b) variations of stress in Si in crystallography system vs depth below the gate; (c) variations of stress in Si in sample system vs depth below the gate and (d) simulation of stress variations in sample system vs depth of the gate (red: Sxx, blue: Syy, green: Szz)

As seen in Table1, Chi^2 are all small (<5) for each fitting, indicating reliable experimental results. Fig. 3 shows (a) variations of Si lattice parameter vs. different depth below the gate and (b) variations of strain and stress in sample system vs. different depth below the gate and (d) simulation using Ge concentration of 15%. Si lattice parameters variations represent tetragonal distortion in the Si crystal with contraction in the <100> ("a") and <010> ("b") directions and expansion in <001> ("c") direction. The strain and stress values rotated from crystal system to sample system show that the Si lattice starts to show compression in the <011> direction and tension <001> direction at ~100nm below the gate, i.e. Poisson effects. The compressive stress reaches a maximum of ~824 MPa close to the surface (~25nm below the gate) where the eSiGe has its greatest effect. The simulated stress value at the same depth below the gate (~25nm below the gate) from Fig. 3(d) is ~925 MPa. The experimental results of the compressive stress of ~824 MPa close to the gate matches with the simulated stress value which was calculated based on the same Ge concentration and critical dimensions of eSiGe device structures. There is a slight difference between the experimental results and the simulated because that the effect of tensile stress induced into the channel area from the poly-crystalline gate is not taken into consideration in the simulation.

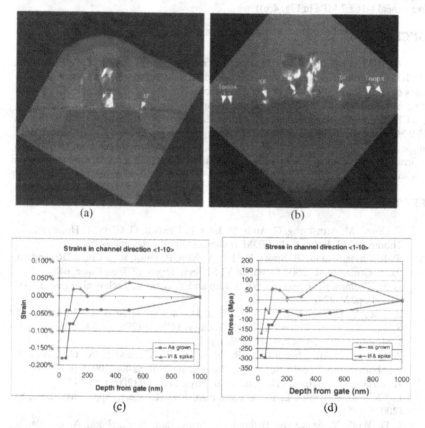

(a) (b)

(c) (d)

Figure 4 Weak beam TEM images of PFET gate with 40nm eSiGe with 15 Ge at%: (a) as grown and (b) after B implanted and spike anneal; (c) and (d): strain and stress along the channel direction (sample system) for these two samples measured by CBED.

Many factors affect stress in channel region. On the positive side, deeper eSiGe recess and higher Ge concentration leads to higher stress. On the negative side, defects within eSiGe reduce stress in channel area. Fig.4(a) and 4(b) shows two TEM images of PFET gates with different depths of eSiGe and different processing conditions and the corresponding strain and stress within channel area ((c) and (d)). Defects within eSiGe film relax the strain/stress within itself and eventually reduce strain/stress in the channel area. After SiGe growth, there is no misfit dislocation seen at the eSiGe/Si interface and only a few small stacking faults are seen at the corner of the capped Si over the eSiGe or, as shown in Fig.4(a). However, after B implant and spike anneal, dislocation loops formed within eSiGe, as shown in Fig 4(b). As a result of the combined effect of deeper eSiGe and absence of defects (Fig. 4(a)), the stress in channel area is approximately twice as high (-285.8 MPa) for the as grown case than that for the sample after B implant and spike anneal (-168.7 MPa in Fig. 4(b))

CONCLUSIONS

CBED has been successfully applied to measure strain/stress in the channel area of PMOS semiconductor devices with embedded SiGe in the source/drain areas. Reliable results of strain/stress measurements within the channel area have been achieved by good fitting of experimental CBED patterns into theoretical calculations. Stress measurements from CBED are in good agreement with stress simulations. Compressive stress as high as 823.9 MPa have been measured by CBED along the <110> direction for an eSiGe PMOS sample with 15% Ge and a thickness of 80nm. Stress measurements from CBED also confirm that the depth of eSiGe and defects within the eSiGe relax strain/stress within itself and reduce strain/stress in the channel area.

REFERENCES

1. T. Ghani, M. Armstrong, C. Auth, M. Bost, P. Charvat, G. Glass, T. Hoffmann, K. Thompson, M. Bohr et, al., IEDM Technical Digest 2003, p978-980
2. P. R. Chidambaram, B. Smith, L. Hall, Y. Kim, P. Jones, R. Irwin, A. Rotondaro, D. T. Grider, et al., 2004 Symp. on VLSI Tech. Diges. of Tech Paper, p48
3. S. E. Thompson, G. Sun, K. Wu, J. Lim and T. Nishida, Technical Digest, IEDM (2004), p221-224
4. Z. Lou, Y. Chong, J. Kim, N. Rovedo, D. Chidambarrao, J. Li, R. Davis, D. Schepis, H. Ng, K. Rim et, al., IEDM Technical Digest (2005), p495-498
5. J. M. Zuo, Ultramicroscopy, 41 (1992), p211
6. A. Armigliato, R. Balboni, A. Benedetti, G. P. Carnevale, A. G. Cullis, S. Frabboni and D. Piccolo, Solid State Phenomena, Vol. 82-84 (2002), p727-734
7. A. Toda, N. Ikarashi, and H. Ono, Appl. Phys. Lett., 79 (2001), p 4243
8. L. Clement, R. Pantel, L. F. Tz. Kwakman, J. L. Rouviere, Appl. Phys. Lett., 85 (2004), p651
9. I. D. Wolf, V. Senez, R. Balboni, A. Armigliato, S. Frabboni, A. Cedola, S. Lagomarsino, Microelectronic Engineering, 70 (2003), 425-435
10. A. Benedetti, A. G. Gullis, A. Armigliato, R. Balboni, S. Frabboni, G. F. Mastracchio, G. Pavia, Appl. Surf. Sci., 188(2002), p214
11. R. Balboni, S. Frabboni and A. Armigliato, Phil. Mag., A, Vol 77(1998), p67

Mater. Res. Soc. Symp. Proc. Vol. 913 © 2006 Materials Research Society   0913-D05-04

A Physically Based Quantum Correction Model for DG MOSFETs

Markus Karner, Martin Wagner, Tibor Grasser, and Hans Kosina
Institute for Microelectronics, TU - Wien, Gußhausstraße 27-29 / E360, Wien, Austria

ABSTRACT

In this work we present a physically based quantum correction model for highly scaled double gate (DG) CMOS devices. In contrast to previous work, our quantum correction model is based on the bound states that form in the Si film. The Eigenenergies and expansion coefficients of the wave functions are tabulated for arbitrary parabolic approximations of the potential in the quantum well. This enables a highly efficient use in TCAD applications.

INTRODUCTION

Due to the strong impact of quantum mechanical effects on the characteristics of today's semiconductor devices, purely classical device simulation without quantum correction is no longer sufficient to provide accurate results. Besides tunnelling, the effect of quantum confinement strongly affects the device characteristics of bulk, silicon-on-insulator (SOI), and double gate (DG) MOSFETs under inversion conditions. Due to quantum confinement, the local density of states is affected and the carrier concentration near the gate oxide decreases, while classical simulation predicts an exponential growth near the gate oxide.

Schrödinger Poisson (SP) solvers, delivering a self consistent solution of the quantum mechanical carrier concentration and the Poisson equation, accurately determine quantum confinement, hence accurately deliver the inversion charge [6], but they are computationally demanding. In order to obtain proper results at significantly reduced CPU time, several quantum correction models for classical simulations have been proposed [1-5]. However, some of these corrections are based on empirical fits with numerous parameters [3,4]. In some other models, the dependence on the electrical field adversely affects the convergence behavior [2]. Practically, the model proposed in [1] has to be recalibrated for each device. A comprehensive comparison of these models can be found in [5]. In addition, none of these models is suitable for highly scaled DG MOSFETs in the deca nanometer regime where two coupled inversion regions occur. In this work, we present a new, physically based, and more specific approach for DG MOSFETs.

APPROACH

The value of the classical carrier concentration, based on the assumption of free 3-dimensional electron gas and Boltzmann statistics, is adjusted to be equal to the quantum mechanical carrier concentration, given by a summation over all contributing subbands, by introducing the quantum correction potential φ_{corr} as

$$n_{cl,corr} = N_C \exp\left(-\frac{E_C - q\varphi_{corr} - E_f}{k_B T}\right),$$

(1)

$$n_{qm} = N_{C1} \sum_n |\Psi_n(x)|^2 \exp\left(-\frac{E_n - E_f}{k_B T}\right). \tag{2}$$

Here, N_C and N_{C1} denote the effective density of states for the classical and quantum mechanical carrier concentration, respectively, φ_{corr} the quantum correction potential, E_C the conduction band edge energy, and E_f the Fermi energy.

This approach requires the knowledge of the energy levels E_n and the wave functions $\Psi_n(x)$ of the quantized states. To avoid the computationally expensive solution of the Schrödinger equation, we tabulate the solutions for a parabolic shaped conduction band edge

$$E_C(x) = E_{max} - a\left(\frac{d}{2} - x\right)^2 \tag{3}$$

as displayed in Fig. 1. Input parameters are the film thickness d and the curvature a which is derived from an initial classical simulation. The wave functions are expanded as

$$\Psi_n(x) = \sum_k \xi_{n,k} \sqrt{\frac{2}{d}} \sin\left(\frac{\pi}{d} kx\right). \tag{4}$$

The offset of the energy levels ε_n and the expansion coefficients of the wavefunctions $\xi_{n,k}$ found are by interpolation of tabulated values. This allows one to estimate a correction potential φ_{corr} such that the corrected classical carrier concentration is consistent with the SP solution

$$\exp\left(-\frac{q\varphi_{corr}}{k_B T}\right) = \exp\left(-\frac{a\left(\frac{d}{2} - x\right)^2}{k_B T}\right) \times \sum_m \frac{N_{C1,m}}{N_C} \sum_n |\Psi_{m,n}(x)|^2 \exp\left(-\frac{\varepsilon_{m,n} - E_f}{k_B T}\right). \tag{5}$$

Here, m denotes the summation over the different valley sorts (three for silicon) [7] and a the parabolic coefficient.

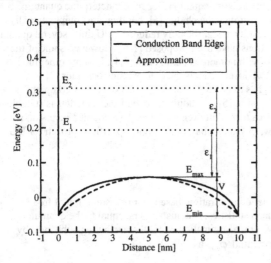

Fig. 1: The conduction band edge in the device structure and its parabolic approximation. Furthermore, the energy level with respect to E_{max} and E_{min} are shown.

RESULTS

We implemented this model in the general purpose device simulator Minimos-NT [8]. Our SP simulator VSP was used to derive the reference QM curves. Fig. 2 shows the electron concentration at different bias points for DG MOSFETs with 5 nm and 10 nm film thickness. Outstanding agreement between the QM and the corrected classical curves (DGTab) is achieved. Since the inversion charge is calculated from the accurate carrier concentration, no further fitting is necessary. Both, the inversion charge and the resulting gate capacitance as a function of the gate bias are shown in Fig. 3. An excellent agreement for a wide range of gate voltages and relevant film thicknesses has been achieved.

CONCLUSIONS

We derived a physically based quantum correction model which accurately reproduces both carrier concentrations and gate capacitance characteristics specifically for DG MOSFETs. Due to its computational efficiency the model is well suited for TCAD simulation environments.

ACKNOWLEDGMENTS

This work has been partly supported by the European Commission, project SINANO, IST 506844 and from the Austrian Science Fund, special research project IR-ON (F2509-N08).

REFERENCES

1. W. Hänsch et al., SSE, vol.32, pp. 839-849, (1991)
2. M. Van Dort et al., SSE, vol. 37, no. 3, pp. 411-414 (1994)
3. C. Jungemann et al., Proc. MSM, pp. 458-461 (2001)
4. C. Nguyen et al., Proc. NSTI-Nanotech, vol. 3, pp. 33-36 (2005)
5. M. Wagner et al., Proc. IWPSD, vol. 1, 458-461 (2005)
6. N. Yang. Et al, IEEE Transaction on Electron Devices, vol. 46, pp. 1464–1471 (1999)
7. F. Stern, Phys.Rev.B, vol. 5, pp. 4891-4899 (1972)
8. IuE, Minimos-NT 2.1 User's Guide, TU Wien,
   http://www.iue.tuwien.ac.at/software/minimos-nt (2004

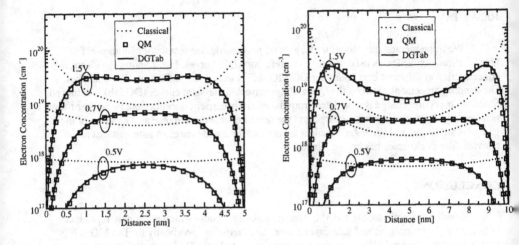

Fig. 2: The classical, the quantum mechanical, and the corrected classical electron concentration of a DG-MOS structure with 10 nm / 5 nm Si-film thickness is shown in the right / left part of the figure respectively.

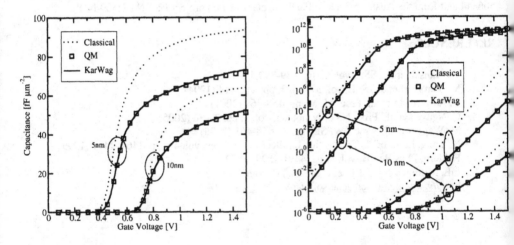

Fig. 3: The total amount of the inversion charge as a function of the gate bias is shown in the right part and the resulting capacitance characteristics is shown in the left part.

Mater. Res. Soc. Symp. Proc. Vol. 913 © 2006 Materials Research Society   0913-D05-05

TCAD Modeling of Strain-Engineered MOSFETs

Lee Smith

Synopsys, Inc., 700 E. Middlefield Road, M/S A14, Mountain View, CA, 94043

ABSTRACT

The rapid rise of standby power in nanoscale MOSFETs is slowing classical scaling and threatening to derail continued improvements in MOSFET performance. Strain-enhancement of carrier transport in the MOSFET channel has emerged as a particularly effective approach to enable significant performance improvements at similar off-state leakage. In this paper we describe how strain effects are modeled within the context of TCAD process and device simulation. We also use TCAD simulations to review some of the common approaches to engineer strain in MOSFETs and to explain how strain impacts device and circuit characteristics.

INTRODUCTION

Strain engineering is rapidly becoming a ubiquitous element in modern MOSFET design [1-3]. In the current stage of technology development, much effort is being spent to model strain effects and to optimize the strain induced by various strain sources. Technology Computer Aided Design (TCAD) tools provide a convenient means of simulating the stress and strain produced during the strain-engineered process flow as well as the impact of that strain on device performance. As will be shown by example, the resulting stress and strain fields are often non-intuitive. The impact of strain on device characteristics is determined primarily through changes in the band structure. In this paper, we review how subsequent changes in carrier repopulation, effective mass, and scattering enhance, or degrade, the mobility and shift the threshold voltage for various stress configurations. In this context, optimizing the enhancement of the low-field mobility can be viewed as an exercise in band structure engineering. For high-field transport, we use Monte Carlo device simulation to investigate the impact of strain on velocity overshoot and drive current. Beyond the analysis and optimization of strain for a single device lies the next stage in strain engineering: the impact of layout. Due to the large interaction range of stress in CMOS materials, approximately 2 μm, the modeling of isolated devices is not sufficient to predict final circuit behavior. In this paper, we also review some simulation studies we have performed to investigate the impact of circuit layout on channel stress and circuit performance.

EXAMPLES OF STRAIN ENGINEERING

Strain can be engineered into a conventional MOSFET structure in many different ways. These different approaches are typically categorized as either global or a local in nature. Global approaches, such as strained-Si on relaxed SiGe, attempt to induce uniform strain throughout the

Si layer across the entire wafer [4]. While this type of approach has received a great deal of academic interest, the first approaches used in production have been of the local type [1]. Local, or process-induced, strain engineering focuses on producing stress in the channel of a single device. Figure 1 shows two of the most commonly used local approaches: a strained nitride capping layer and embedded SiGe (e-SiGe) in the source/drain regions. The stress from these approaches can be calculated throughout the process flow by solving the force equilibrium equations while considering various sources of stress such as intrinsic film stress, thermal mismatch, and lattice mismatch. Often, the resulting stress fields are non-intuitive.

For example, simulations of tensile strained nitride cap layers for NMOS consistently predict that the vertical stress component is the dominant stress in the channel, rather than the longitudinal stress component, as is commonly assumed. This vertical stress is produced by the tensile cap layer pushing the gate stack down onto the silicon substrate. The longitudinal stress is produced by the cap layer pulling the edges of the gate stack away from the channel.

In the case of e-SiGe source/drain, the induced stress is primarily uniaxial along the longitudinal, or channel, direction. The embedded $Si_{1-x}Ge_x$ regions are fabricated by first etching a recess in the silicon substrate and then growing $Si_{1-x}Ge_x$ via selective epitaxy. Due to the lattice mismatch between $Si_{1-x}Ge_x$ and Si, the compressively strained $Si_{1-x}Ge_x$ pushes out against either end of the channel inducing a compressive, longitudinal channel stress. The stress obtained in the channel depends on many properties of the SiGe regions such as the Ge mole fraction, the SiGe recess depth and elevation height, and the shape of the SiGe regions near the channel [3].

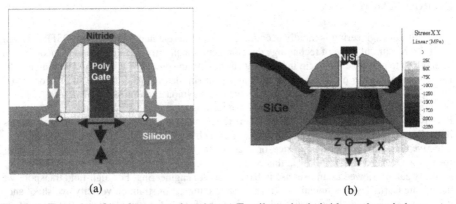

(a) (b)

Figure 1. Examples of local stress engineering. a) Tensile strained nitride cap layer induces longitudinal tension and vertical compression into the channel. b) e-SiGe source/drain regions induce longitudinal compressive stress in the channel.

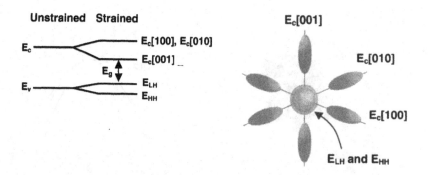

Figure 2. Band splitting in Si due to uniaxial compressive stress along [001].

MODELING THE IMPACT OF STRESS ON DEVICE BEHAVIOR

Stress modifies many aspects of device behavior such as the leakage current, threshold voltage, and mobility. The primary link between stress and these changes to device behavior is through the band structure. As depicted in Figure 2, stress, in general, splits and shifts the silicon band extrema. In the example shown here, uniaxial compressive stress along [001] splits the six Δ valleys of the conduction band into two different groups. Likewise, the heavy and light hole bands are split with the light hole band moving up in energy. These shifts cause a change in the band gap and electron affinity and a subsequent change in threshold voltage. In this example, the band gap is reduced which can lead to an increase in leakage current through increased recombination current and band-to-band tunneling.

Band splitting due to stress also alters valley repopulation and inter-valley scattering and, therefore, the mobility. Under most types of stress, the conduction valleys in silicon undergo rigid shifts with negligible change in effective mass. Recent work, however, suggests the electron effective masses can also be modified under appropriate stress [5]. Because the relaxed silicon valence band consists of degenerate valleys with a high degree of band warping, hole transport in silicon is particularly sensitive to stress. In addition to level splitting, stress significantly modulates the band curvature and the hole effective masses.

Physical model for pmos mobility

A physical approach to modeling the stress dependence of mobility starts with a detailed calculation of the band structure under stress. For example, Figure 3 shows the valence band structure of relaxed and strained Si under uniaxial, compressive stress along [110] computed using a 6-band **k·p** approach [6]. The top valence band in relaxed Si is very warped, and the band dispersion along [110] and [001] shows the familiar degenerate light hole and heavy hole bands at the Γ point. Under uniaxial [110] compressive stress, the light and heavy hole bands

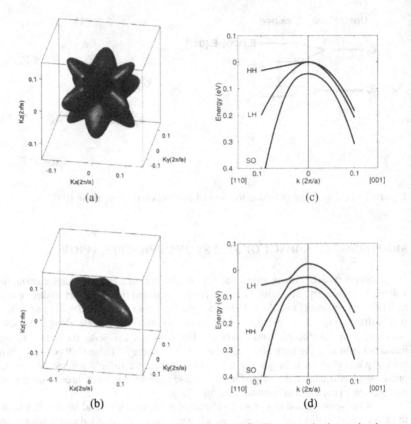

Figure 3. Bulk valence band structure for relaxed and strained Si. The stress in the strained case is 1 GPa of uniaxial, compressive stress along [110]. Figures are: Iso-energy contours throughout k-space at 50 mV below the top of the band for a) relaxed Si and b) strained Si; band dispersion along [110] and [001] for c) relaxed Si and d) strained Si.

split, and the top band simplifies to an almost single ellipsoidal band with one arm oriented along [-110]. The band dispersion along [110], the typical MOSFET channel direction, places the light hole band as the top band with a reduced mass along [110].

While mobility can be computed rigorously from the band structure and scattering rates, a simplified model often provides good accuracy and a more intuitive understanding of the important mechanisms involved [7]. As shown in Figure 4, the stress-induced changes to the top-most valence band are modeled using two ellipsoidal valleys oriented along [110] and [-110].

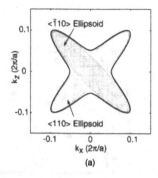 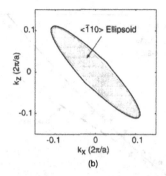

(a)　　　　　　　　　　　　(b)

Figure 4. Thick lines show the constant energy contour of the top-most valence band at 50 mV below the top of the band for (a) relaxed and (b) uniaxially stressed Si. Also sketched are the ellipsoidal valleys used in the mobility model.

Each valley is characterized by a transverse and longitudinal effective mass. Assuming a constant mean scattering time $\langle \tau \rangle$, the mobility along the [110] direction can be modeled as:

$$\mu_{[110]} = q \langle \tau \rangle \left(\frac{f}{m'_{[-110]}} + \frac{1-f}{m'_{[110]}} \right) \tag{1}$$

where f is the occupancy of the [-110] valley, $m'_{[-110]}$ is the transverse mass of the [-110] valley, $m'_{[110]}$ is the longitudinal mass of the [110] valley and q is the electron charge. The stress-dependence of the energy level splitting between the two valleys and the effective masses are expanded as polynomials of elements of the stress tensor.

The mobility model is calibrated against measured data from wafer-bending experiments as well as from strain-engineered PMOSFETs that employ various combinations of e-SiGe source/drain and strain capping layers implemented as a compressive etch stop layer (CESL) [8, 9]. For the strain-engineered MOSFETs, the shape of the e-SiGe source/drain and the properties of the stressed cap layer were engineered to vary the compressive stress in the channel from 200 MPa to 2.0 GPa. Figure 5 shows the measured and modeled mobility gain as a function of the effective channel stress. For wafer-bending data, the channel stress is obtained directly from curvature measurements. For the strain-engineered MOSFETs, the channel stress is calculated using 2D process simulation where all intentional and unintentional stress sources as well as the stress evolution during the entire process flow are taken into account [10]. While the channel stress is primarily uniaxial along the channel, there is a non-negligible transverse component that degrades the mobility. This is treated here via an effective channel stress that represents the equivalent reduction in longitudinal stress.

The model agrees well with both the wafer-bending data and the mobility gain extracted from the strain-engineered MOSFETs. Compared to the bulk piezoresistance model [11], the

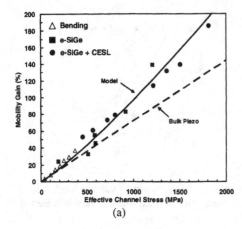

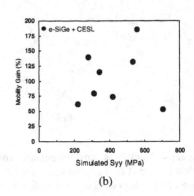

(a)   (b)

Figure 5. a) Measured and modeled hole mobility enhancement as a function of the effective channel stress. The measured data are shown as symbols, the dashed line shows bulk piezoresistance, and the solid line shows the mobility model described in the text. The bending data are from [12]. b) Measured mobility gain as a function of simulated vertical stress showing no correlation.

actual response is moderately superlinear at high stress. The stress dependence of the effective masses and band splitting that result from the calibrated fit indicate that the mobility enhancement arises from two source: repopulation of holes into the top [-110] ellipsoidal valley that has a small transverse mass along [110], and a reduction of this transverse mass as stress is increased. An almost 3x enhancement in the mobility is obtained at 2 GPa of channel stress. Importantly, no saturation in the mobility enhancement is seen even at this stress level. While it is expected that the valley occupancy and transverse mass improvement will saturate at high level splitting, further improvements in the mobility can come from the suppression of scattering.

While the mobility enhancement for vertical stress cannot be characterized via wafer-bending on planar devices, the combination of measured mobility gain and simulated stress allows for an indirect characterization of this stress component. The measured mobility gain as a function of the simulated vertical stress is shown in Figure 5b. No correlation between the mobility gain and the vertical stress is seen, indicating that the effective piezoresistance coefficient for vertical stress is small in (100) PMOS, in agreement with the bulk piezoresistance model.

Stress-enhanced mobility and effective field

When dealing with stress-enhanced mobility in MOSFETs an important consideration is the dependence of the mobility enhancement on vertical effective field. Device scaling continues to increase the effective field with values reaching over 1.5 MV/cm at the 65 nm technology

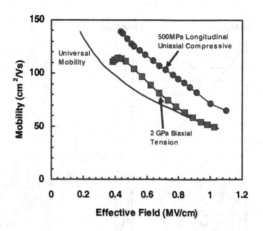

Figure 6. Effective low-field channel mobility as a function of vertical effective field for: the Si universal mobility curve [13], biaxial in-plane tension [2], and uniaxial longitudinal compression [1].

node. One view of stress enhancement techniques is that they allow the "tyranny of the Si universal mobility curve" to be broken. The goal is to find stress tensors that are able to shift the entire universal mobility curve to a higher level over the entire effective field range. Figure 6 compares the effective field dependence of the mobility enhancement for two different types of stress tensors in PMOS: biaxial, in-plane tension and uniaxial, longitudinal compression. While biaxial tension produces a healthy enhancement at low effective field, the enhancement is lost as the effective field is increased. In contrast, uniaxial, longitudinal compression is able to maintain a large mobility enhancement over the entire effective field range.

The difference in behavior between these two stress cases lies in how the subband structure changes under stress. Figure 7 shows the top two subbands for these two stress cases at low and high effective field computed using a triangular well approximation within the 6-band k·p approach [14]. At low effective field, both stress cases produce a healthy separation between the top two subbands which reduces inter-subband scattering and maintains a small effective mass along the transport direction. At high effective field, however, the separation between the top subbands for the biaxial case is reduced as the effects of band splitting due to stress and quantization work against each other. This leads to increased inter-subband scattering and reduced mobility. In addition to subband splitting, the effective masses at the top of the subbands along the transport direction are affected differently by the two stress cases. At high effective field, uniaxial stress produces a small effective mass while biaxial stress produces a large mass which is detrimental to mobility.

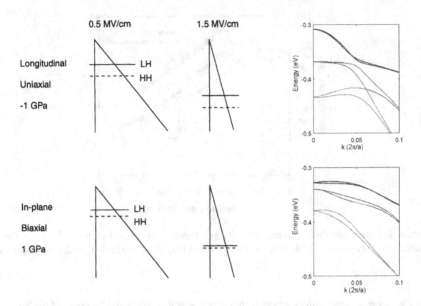

Figure 7. Comparison of subband behavior under uniaxial and biaxial stress in PMOS under low and high effective field. The band dispersion shows the top six subbands along [110] at high effective field. The biaxial stress case produces a large effective mass in the top subband which is detrimental to mobility.

High-field transport under stress

While enhancement of the low-field mobility is an important objective for stress engineering, the ultimate goal is to enhance the final device performance in terms of improved drive current. In addition to increased vertical effective field, device scaling also increases the longitudinal driving field from source to drain. This pushes devices further into the quasi-ballistic regime. Competing effects such as velocity saturation and velocity overshoot now become important in determining the final drive current and how the low-field mobility enhancement is translated into drive current enhancement.

Monte Carlo device simulation provides a direct means of investigating the impact of stress on high field transport. Both Monte Carlo simulations and experiments of high-field transport in strained Si show that the saturation velocity is not significantly enhanced by stress. In contrast, simulations suggest that velocity overshoot is strongly affected. Figure 8 compares the simulated transient velocity overshoot effect in bulk Si for relaxed and biaxally strained Si with a stress equivalent to a strained-Si layer grown on relaxed $Si_{0.8}Ge_{0.2}$. Repopulation of carriers into valleys with a small effective mass along the transport direction leads to significant enhancement of the velocity in the overshoot regime. The simulation of a 25 nm gate length NMOSFET shows a drive current enhancement of 30% as compared to a 75% enhancement of

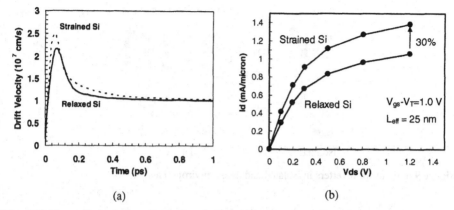

(a)   (b)

Figure 8. Monte Carlo device simulation of relaxed Si and strained Si on $Si_{0.8}Ge_{0.2}$. a) Comparison of transient velocity overshoot behavior in n-type Si. b) Comparison of drain characteristics for a 25 nm gate-length NMOSFET.

the low-field mobility. This ratio of approximately 0.5 between drive-current gain and the low-field mobility gain has also been seen experimentally [15]. We have extracted a similar ratio between the drive-current gain and the low-field mobility gain for PMOS under uniaxial stress [9].

MODELING THE IMPACT OF LAYOUT ON STRESS AND CIRCUIT BEHAVIOR

While local stress sources such as strained capping layers and e-SiGe source/drain are optimized to induced stress into the channel of a single MOSFET, the stress fields that they produce can extend quite a distance beyond the target transistor. Even unintentional stress sources such as STI can induce significant stress over a 2 µm range. These stress proximity effects (SPE) make the final stress in the channel of a MOSFET depend on the environment in which the MOSFET is placed, i.e. on the circuit layout.

As an example of SPE we consider the impact of layout density on the performance of inverters in a ring oscillator. As diagrammed in Figure 9, in a sparse layout a single, isolated transistor is surrounded by a large area of STI. In a dense layout, transistors are closely nested and separated by a small amount of STI. Detailed analysis of the stress fields produced by these layouts was performed using 3D process simulation [16].

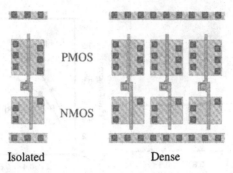

PMOS

NMOS

Isolated   Dense

Figure 9. Layout for inverters in isolated and dense environments.

    We first consider MOSFETs without any engineered stress sources. In this case the STI produces unintentional biaxial compressive stress in the isolated transistors. This has little effect on the PMOSFET mobility but degrades the performance of the NMOSFET. In the dense layout the small amount of STI between the transistors is unable to generate significant stress and, therefore, the longitudinal stress component is suppressed. This is beneficial for the NMOSFET mobility but detrimental for the PMOSFET performance. The net impact of these layout-induced changes to stress on circuit performance was investigated by simulating the response of a 3-stage ring oscillator. The device output characteristics shown in Figure 10 were computed using energy balance device simulation to consider the impact of stress on both the low-field mobility and velocity overshoot. Overall, the PMOSFET response to layout determines the overall change in circuit response, causing the oscillator frequency to slow when changing from a sparse to dense layout.

    Adding an engineered stress source to the PMOSFET completely changes the layout dependence. Including e-SiGe source/drain into the PMOSFET induces beneficial stress into the channel. The amount of channel stress that can be generated, however, depends on the environment around the e-SiGe stressor. Because of the relative softness of STI, a large area of STI can act as a stress relaxor and reduce the amount of channel stress obtained. The effectiveness of the e-SiGe stressor is therefore degraded in the sparse layout. In the dense layout, little stress relaxation occurs because of the limited area of STI along the longitudinal direction. The change in oscillator performance between these two layouts is shown in Figure 10. The enhanced PMOSFET performance in the dense layout now greatly improves the overall oscillator speed as compared to the sparse layout.

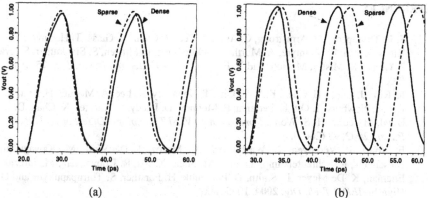

(a)   (b)

Figure 10. Comparison of ring oscillator behavior for sparse and dense layouts. a) STI as the only stress source. b) STI and e-SiGe as stress sources.

SUMMARY

Strain engineering is now a critical part of modern MOSFET design. TCAD process simulation provides a convenient means of analyzing the interaction of multiple stress sources during the process flow. The results produced by these complex interactions as well as the relaxation induced by STI are often not intuitive. In terms of device behavior, strain engineering can be viewed as an exercise in band engineering. The change in band structure with strain provides a physical basis for modeling strain-induced mobility enhancement and for identifying particular stress tensors that are beneficial for enhancing device performance. As device scaling continues, strain effects on quasi-ballistic transport will become more important in setting the ratio of drain current enhancement to low-field mobility enhancement. Due to the large range of stress proximity effects in CMOS materials, the modeling of isolated devices is not sufficient to predict final circuit behavior. The impact of circuit layout on channel stress will need to be considered as well.

ACKNOWLEDGMENTS

The author would like to thank Victor Moroz, Xiaopeng Xu, Dipu Pramanik, and Fabian Bufler from Synopsys for useful discussions and simulations and Faran Nouri and Lori Washington from Applied Materials, and Geert Eneman from IMEC, for useful discussions and device data.

177

REFERENCES

1. S. E. Thompson, M. Armstrong, C. Auth, S. Cea, R. Chau, G. Glass, T. Hoffman, J. Klaus, Z. Ma, B. Mcintyre, A. Murthy, B. Obradovic, L. Shifren, S. Sivakumar, S. Tyagi, T. Ghani, K. Mistry, M. Bohr and Y. El-Mansy, *IEEE Electron Device Lett.* **25**, 191 (2004).
2. K. Rim, J. Chu, J.H. Chen, K.A. Jenkins, T. Kanarsky, K. Lee, A. Mocuta, H. Zhu, R. Roy, J. Newbury, J. Ott, K. Petrarca, P. Mooney, D. Lacey, S. Koester, K. Chan, D. Boyd, M. Ieong, H.-S. Wong, *Symposium on VLSI Technology, Digest of Technical Papers* **2002**, 98 (2002).
3. F. Nouri, P. Verheyen, L. Washington, V. Moroz, I. De Wolf, M. Kawaguchi, S. Biesemans, R. Schreutelkamp, Y. Kim, M. Shen, X. Xu, R. Rooyackers, M. Jurczak, G. Eneman, K. De Meyer, L. Smith, D. Pramanik, H. Forstner, S. Thirupapuliyur and G. S. Higashi, *IEDM Tech. Dig.* **2004**, 1055 (2004).
4. J. Welser, J.L. Hoyt and J.F. Gibbons, *IEEE Electron Device Lett.* **15**, 100 (1994).
5. K. Uchida, T. Krishnamohan, K.C. Saraswat and Y. Nishi, *IEDM Tech. Dig.* **2005**, 129 (2005).
6. T. Manku and A. Nathan, *J. Appl. Phys.* **73**, 1205 (1993).
7. B. Obradovic, P. Matagne, L. Shifren, X. Wang, M. Stettler, J. He and M. D. Giles, *IWCE Tech. Dig.*, 26 (2004).
8. L. Smith, V. Moroz, G. Eneman, P. Verheyen, F. Nouri, L. Washington, M. Jurczak, O. Penzin, D. Pramanik and K. De Meyer, *IEEE Electron Device Lett.* **26**, 652 (2005).
9. L. Washington, F. Nouri, S. Thirupapuliyur, G. Eneman, P. Verheyen, V. Moroz, L. Smith, X. Xu, M. Kawaguchi, T. Huang, K. Ahmed, M. Balseanu, L. Xia, M. Shen, Y. Kim, R. Rooyackers, K. De Meyer and R. Schreutelkamp, *IEEE Electron Device Lett.*, accepted for publication (2006).
10. V. Moroz, N. Strecker, X. Xu, L. Smith and I. Bork, *Material Science in Semiconductor Processing* **6**, 27 (2003).
11. C. S. Smith, *Phys. Rev.* **94**, 42 (1954).
12. E. Wang, P. Matagne, L. Shifren, B. Obradovic, R. Kotlyar, S. Cea, J. He, Z. Ma, R. Nagisetty, S. Tyagi, M. Stettler and M. D. Giles, *IEDM Tech. Dig.* **2004**, 147 (2004).
13. S. Takagi, A. Toriumi, M. Iwase and H. Tango, *IEEE Trans. on Electron Devices* **41**, 2357 (1994).
14. S. Rodriguez, J. A. Lopez-Villanueva, I. Melchor and J. E. Carceller, *J. Appl. Phys.* **86**, 438 (1999).
15. A. Lochtefeld and D. Antoniadis, *IEEE Electron Device Lett.* **22**, 591 (2001).
16. V. Moroz, G. Eneman, P. Verheyen, F. Nouri, L. Washington, L. Smith, M. Jurczak, D. Pramanik and X. Xu, *International Conference on the Simulation of Semiconductor Processes and Devices (SISPAD)* **2005**, 143 (2005).

Mater. Res. Soc. Symp. Proc. Vol. 913 © 2006 Materials Research Society   0913-D05-07

Predictive Model for B Diffusion in Strained SiGe Based on Atomistic Calculations

Chihak Ahn[1], Jakyoung Song[2], and Scott T. Dunham[1,2]

[1]Physics, University of Washington, Seattle, WA, 98195

[2]Electrical Engineering, University of Washington, Seattle, WA, 98195

ABSTRACT

Using an extensive series of first principles calculations, we have developed general models for the change in energy of boron migration state via interstitial mechanism as a function of local alloy configuration. The model is based on consideration of global strain compensation as well as local effects due to nearby arrangement of Ge atoms. We have performed KLMC (Kinetic Lattice Monte Carlo) simulations based on change in migration energy to explain the reduced B diffusion in strained SiGe and compared our results to experimental observations. These models include significant effects due to both global stress and local chemical effects, and accurately predict the B diffusivity measured experimentally in strained SiGe on Si as a function of Ge content.

INTRODUCTION

There is great interest in utilizing SiGe for enhanced mobility, increased activation, and reduced contact resistance. In order to control device structures at the nanoscale, a fundamental understanding of the effects of alloy concentration and associated strain is critical. Even though there have been many experiments showing that boron diffusion is retarded in strained SiGe, the physical mechanism is not well understood. Kuo et al. concluded that stress effects are not significant [1], but Zangenberg et al. reported very strong stress dependence of B diffusivity [2]. And, although Lever et al. attributed diffusivity reduction to B-Ge pairing [3], Hattendorf et al. found that there is no significant binding between B and Ge using β-NMR technique [4]. Previous ab-initio calculations by Wang et al. suggested that the presence of Ge increases the migration energy and reduces concentration of Si interstitials [5]. In this paper we investigated B diffusion mechanism in strained SiGe to solve the controversy, considering both global strain compensation and local Ge configuration.

SIMULATION AND RESULTS

B diffuses in Si lattice mainly via interstitial mechanism [6], and previous research indicates that boron migration occurs via a two step process: from substitutional B with neighboring tetrahedral Si ($B_sSi_i^T$) to B in one of 6 hexagonal sites (the subset of 12 hexagonal sites away from the given Si_i^T site) and then back to one of 6 substitutional sites [7,8]. The B transition state is located between substitutional site and hex site along <311> direction. We calculated the change in interstitial-mediated boron transition state energy in 64 (or 65 with I) atom supercell subject to periodic boundary condition under various Ge configurations, using

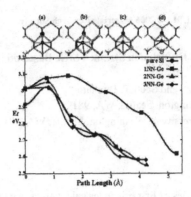

Figure 1. The energy along one step of boron diffusion path ($B_i^{hex} \Rightarrow B_sSi_i^T$) in pure Si and $Si_{63}Ge$. The highest barrier is for Ge in the first nearest neighbor (1NN) site to final B_s (square symbols), followed by Ge at 2NN (triangles). Note that although energy of transition state for Ge in 3NN site is almost the same as for pure Si, a higher barrier would have been required for B to have initially come from any substitutional site in hexagonal ring other than the final site in its previous hop.

Table I : The formation energy difference of the transition state for B diffusion in $Si_{63}Ge$ relative to pure silicon. -1NN indicates Ge nearest neighbor to initial B_s site, but outside of 6-membered ring surrounding final hexagonal site.

One Ge-atom	1NN	2NN	3NN	-1NN
$\Delta E(n^{th})$	0.099	0.047	0.020	-0.027

Table II : The formation energy difference of the transition state for B diffusion in $Si_{62}Ge_2$ relative to pure silicon. Note that increase in energy of transition state is greatest for Ge present on opposite sides of ring, with strongest effect for 1NN.

Two-Ge atoms	1NN-1NN	1NN-2NN	1NN-2NN (Ge-Ge bond)	1NN-3NN	2NN-2NN	2NN-3NN
ΔE_{ij}	0.48	0.25	0.16	0.14	0.12	-0.003

density functional theory (DFT) in generalized gradient approximation (GGA) with ultrasoft Vanderbilt pseudopotentials [9]. The VASP software [10] was used with 340 eV energy cutoffs and 2^3 Monkhorst-Pack k-point sampling. The diffusion paths were calculated using the nudged elastic band (NEB) method [11].

We found the B transition state energy to depend significantly on the local Ge configuration in surrounding hexagonal ring. Fig. 1 shows the energy along the boron diffusion path in pure Si and $Si_{63}Ge$. The formation energy of the transition state is given by

$$E_f = E^{TS}(BSi_{64-x}Ge_x) - xE(Si_{63}Ge) - E(BSi_{63}) + E(Si_{64})[(63+63x-64+x)/64]. \quad (1)$$

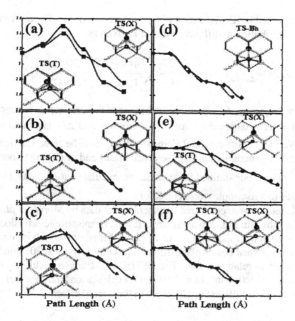

Figure 2. The boron diffusion paths in $Si_{62}Ge_2$ for possible Ge configurations on 6-membered ring. The solid lines and dashed lines are for the diffusion paths to $B_sSi_i^T$ and split interstitial sites, respectively. As the Ge content increases, the split interstitial structures can have lower energy than $B_sSi_i^T$ for some Ge arrangements. With two Ge-atoms, the B diffusion path to the substitutional site with two 1NN Ge-atoms, as shown in Fig. 2(a), has the highest activation energy. The paths for the 1NN and 2NN with/without Ge-Ge bond are shown in Fig. 2(b) and (c). The boron diffusion in a ring with Ge-Ge bond has lower activation energy as the B_i is apparently more able to keep away from Ge atoms.

Using calculations over a wide range of Ge configurations, we extracted a discrete model for the energy of B transition state as a function of local Ge arrangement. To analyze the local Ge effects, we separated out the global strain compensation energy ($\Delta E_s = -\Omega_0 \Delta\bar{\varepsilon} \cdot C \cdot \bar{\varepsilon}$) from transition state energy change (Table 1 and 2). Detailed calculations on global strain compensation could be found in previous work [12,13]. After we separated out the effect of global strain compensation, we found that the dominant effect is for Ge within the 6-membered ring surrounding target hexagonal site. The model is parameterized in terms of the number of Ge atoms at 1st, 2nd and 3rd NN distances from terminal B_s site within the hexagonal ring surrounding the target hexagonal site . The presence of Ge in proximity to the transition state increases the migration energy as seen in Figs. 1 and 2. In the transition states, B tends to shift away from Ge sites due to a repulsive B_i-Ge interaction.

181

The model is used in KLMC under arbitrary stress condition. For a given initial configuration, the contribution to diffusivity tensor is given as,

$$D_{pq}(\bar{\varepsilon})=\sum_{i}\sum_{j}\Gamma_0\exp\left(-\frac{E_i^{m1}(\bar{\varepsilon})}{kT}\right)\frac{\Gamma_0\exp\left(-E_j^{m2}(\bar{\varepsilon})/kT\right)}{\sum_{k}\Gamma_0\exp\left(-E_k^{m2}(\bar{\varepsilon})/kT\right)}\Delta x_{ij}^p\Delta x_{ij}^q \qquad (2)$$

where Γ_0 is attempt frequency, E_i^{m1} is migration barrier of the first hop to hexagonal site, E_j^{m2} is second migration barrier to lattice sites around that hex site, and Δx_{ij}^p is p^{th} component of hopping vector. We find the off-diagonal elements of D_{pq} to all be zero. The diffusivity is statistically sampled by generating random Ge distributions and averaging to obtain the change in diffusivity: $\overline{D}_{pq}^{SiGe}/\overline{D}_{pq}^{Si}$. For strained SiGe on Si, we calculate out-of-plane $\overline{D}_{33}$ and in-plane $\overline{D}_{11}$ diffusivities due to anisotropic stress conditions.

The KLMC results were compared to experimental data by Moriya et al. [14] and Fang et al. [15] (Fig. 3). As shown in Fig. 3, biaxial stress results in anisotropic diffusion. The change in in-plane diffusivity is stronger than that in out-of-plane diffusivity, which is consistent with Diebel et al. [8]. We matched out-of-plane diffusivity with Moriya's at 20% since Moriya's diffusivities were given as relative values. The DFT/KLMC analysis accurately reproduces the extent of B diffusivity reduction in strained SiGe. Our calculations indicate that both stress and local chemical effects are significant contributors to the overall effect.

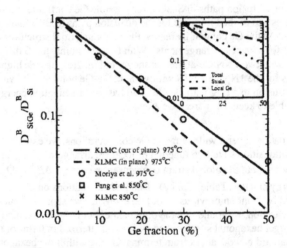

Figure 3. B diffusivity in strained SiGe. Note that appropriate comparison for data is with out-of-plane diffusivity calculation, since diffusion was measured in vertical direction only. Data from Moriya et al. [13] were normalized to 20% Ge result since Moriya reported only relative diffusivity. It can be seen that the calculations do an excellent job of predicting change in B diffusion with Ge fraction. The inset shows strain effects (dotted line) and local Ge effects (broken line) at 975°C separately.

CONCLUSIONS

We analyzed the complicated B diffusion in strained SiGe alloys, using a combination of atomistic models, DFT and KLMC. We found from DFT results that B transition state energy strongly depends on both global strain compensation and local Ge configuration. By separating strain and local Ge effects and performing KLMC, we developed predictive model for retarded B diffusion in strained SiGe. For B diffusion in epitaxial SiGe on Si, strain effects are somewhat stronger than local Ge effects, but both effects are significant.

ACKNOWLEDGMENTS

The authors would like to thank SRC for supporting research and Intel and AMD for donating computer hardware used in this work.

REFERENCES

1. P. Kuo, J. L. Hoyt, J. F. Gibbons, J. E. Turner, and D. Lefforge, Appl. Phys. Lett. **66** (5), 580-582 (1995).
2. N.R. Zangenberg, J. Fage-Pedersen, J. L. Hansen, A.N. Larsen, J. Appl. Phys. **94**, 3883 (2003).
3. R.F. Lever, J.M. Bonar, and A.F.W. Willoughby, J. Appl. Phys. **83**, 1988 (1998).
4. J. Hattendorf, W.-D. Zeitz, W. Schröder, and N.V. Abrosimov, Physica B **340-342**, 858 (2003).
5. L. Wang, P. Clancy, and C.S. Murthy, Phy. Rev. B **70**, 165206 (2004).
6. A. Ural, P.B. Griffin, and J.D. Plummer, J. Appl. Phys. **85**, 6440 (1999).
7. W. Windl, M.M. Bunea, R. Stumpf, S.T. Dunham, and M.P. Masquelier, Phys. Rev. Lett, **83**, 4345 (1999).
8. M. Diebel, Ph.D Thesis, Univ. of Washington (2004).
9. D. Vanderbilt, Physical Review B **41**, 7892 (1990); G. Kresse and J. Hafner, J. Phys: Cond. Matter **6**, 8245 (1994).
10. G. Kresse and J. Hafner, Phys. Rev. B **47**, 558 (1993); **49**, 14251 (1994); G. Kresse and J. Furthmuller, Comput. Mater. Sci. **66**, 16 (1996); Phys. Rev. B **55**, 11169 (1996).
11. G. Henkelman, B. Uberuaga and H. Jónsson, J. Chem. Phys. **113**, 9901 (2000).
12. C. Ahn, M. Diebel, and S.T. Dunham, JVST B **24**, 700 (2006).
13. S.T. Dunham, M. Diebel, C. Ahn, and C.-L. Shih, JVST B **24**, 456 (2006).
14. N. Moriya, L.C. Feldman, H.S. Luftman, C.A. King, J. Bevk, and B. Freer, Phys. Rev. Lett. **71**, 883 (1993).
15. T.T. Fang, W.T.C. Fang, P.B. Griffin, and J.D. Plummer, Appl. Phys. Lett. **68**, 791 (1996).

Mater. Res. Soc. Symp. Proc. Vol. 913 © 2006 Materials Research Society　　　　0913-D05-08

3D Modelling of the Novel Nanoscale Screen-Grid FET

Pei W. Ding[1], Kristel Fobelets[1], and Jesus E. Velazquez-Perez[2]
[1]Department of Electrical Engineering, Imperial College, London, United Kingdom
[2]Departamento de Fisica Aplicada, Universidad de Salamanca, Salamanca, Spain

ABSTRACT

A novel field effect transistor (FET) that uses 3-dimensional (3-D) embedded gate fingers – the Screen-Grid Field Effect Transistor (SGFET) – is proposed. The gating action of the SGFET is based on the design of multiple gating cylinders into the channel region, perpendicular to the current flow. Such configuration allows a full 3-D gate control of the current which improves the device characteristics by increasing the gate to channel coupling. Initial investigations of the SGFET using 3-D TCAD Taurus[TM] simulation software are presented in this paper. The results indicate that the proposed SGFET offers the possibility of downscaling without degrading the output characteristics. A comparison between the SGFET and both bulk and SOI MOSFETs shows the superior characteristics of the SGFET for low power operation.

INTRODUCTION

Progress in semiconductor industry is determined by increased operation speed and packing density, and decreased power consumption which are obtained by means of a reduction of the geometrical parameters. Downscaling of the device dimensions goes hand-in-hand with increased short channel effects [1], such as high off-current and drain induced barrier lowering (DIBL). These can be combated, to a certain extend, via ingenious but expensive fabrication techniques, or via the introduction of novel device geometries. Both innovative device structures (e.g. finFET) and new materials (e.g. strained-Si) have been proposed for such purposes. In this work we propose a novel device geometry based on the principle of embedded gate fingers – the SGFET. The DC operation of the SGFET is investigated via TCAD using the 3-D Taurus[TM] simulation software [2]. The influence of the geometrical position of the gating cylinders and their dimensions on the electrical performance of the device is investigated, and compared to bulk MOSFETs and SOI MOSFETs with the same source to drain distance.

DEVICE STRUCTURE AND OPERATION

The proposed SGFET is based on the design of the gating cylinders perpendicular into the channel region. A particularly promising gating configuration is shown in Figure 1, but other configurations and gate finger geometries are possible. The structure is quite similar in principle to the permeable base transistor (PBT), but with reduced processing complexity, compatible with current CMOS technology [3]. Unlike the PBT, the SGFET is a unipolar device in which the type of the majority carriers is determined by the doping of the ohmic contacts and the channel region. The SGFET has to be fabricated on silicon on insulator (SOI) or strained-silicon on insulator (SSOI) [4]. Cylindrical gate fingers are embedded and defined into the SOI body perpendicular to the current flow, allowing a full 3-D gating control of the current. The gating holes can be defined by e-beam lithography and metal is used for the gate contact, consistent with the prediction in the International Technology Roadmap for Semiconductors (ITRS) [5]. For high mobility operations, the silicon channel can be lowly doped or undoped. A relatively thick insulator is deposited to cover the active area of the SGFET and thus reduce the surface capacitance. This ensures that the gating capacitance is solely related to the gates inside the channel, with surrounding thin gate oxide defined by thermal oxidation. Control of the gate oxide quality is similar to trigate FETs and finFETs with vertical gate sidewall. The cylindrical gate definition into the channel should reach the buried oxide of the SOI whereas the ohmic contacts are introduced as in the current standard MOSFET processing technology.

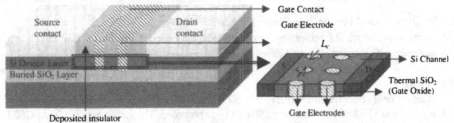

Figure 1. l.h.s. 3-D structure of proposed SGFET. The r.h.s. figure is a detail of the channel layer (Si body) with a 6-hole gate cylinder configuration. L_c is the inter-gate distance that determines device performance. The 2nd row of gate cylinders near the drain controls DIBL.

Figure 2 compares the gating action of a classical MOSFET and the proposed SGFET, where the gating effect of a single embedded gate cylinder is shown. In the SGFET, the electric field lines (full arrows) are pointing radial in all directions from the gate axis along the length of the gate cylinder, covering the whole depth of the channel. The current through the channel is fully controlled by the depletion region extending between the different gate cylinders arranged in an array, similar to MESFET operation but retaining the possibility of enhancement mode operation. This 3-D gating maximises gate control and minimises leakage currents. The efficiency of the gate is determined by the inter-gate distance L_c and not by the source to drain (L_{S-D}) distance. In the case of a classical MOSFET both L_{S-D} and performance are related because when L_{S-D} is reduced to decrease transit time, the gate length has to shorten leading to an increased channel conductance g_d (DIBL) which reduced the low frequency voltage gain. Similar limitations, although to a reduced extend, apply to other viable FET structures including multi-gated FETs. The SGFET can prevent this problem based on the novel gate topology where the gating action is controlled by L_c and is thus decoupled from L_{S-D}.

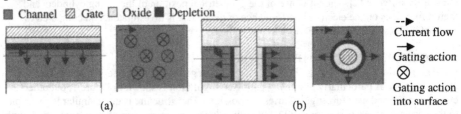

Figure 2. Gating action for side view cross-section (l.h.s.) and top view cross-section of the channel region (r.h.s.) of (a) a classical MOSFET and (b) a SGFET respectively.

DEVICE SIMULATION

Electrical characteristics of the SGFET were studied through extensive 3-D TCAD device simulations using the Taurus[TM] software based on the drift-diffusion model. By solving Poisson's equation and the electron and hole current continuity equations, Taurus[TM] models the

3-D distribution of the potential and carrier concentration in the device to predict its electrical characteristics for any bias condition. To simplify our study, we assumed that source and drain are neither under- nor over-lapped by the gate and that the gate oxide is ideal. The proposed device structure with the side and top cross-section as given in Figure 3, was simulated to study the effects of various geometrical parameters on the device characteristics. The simplified structure – a device unit cell – was used to speed up the 3-D Taurus™ simulations. The actual device is built via a repetition of the device unit cell by translating the unit cell in Figure 3b over the channel width. Simulations confirm that the current, I_{DS}^N of a complete SGFET with $2N$ gate cylinders (in a 2 row configuration for $N=2, 3, 4$ and 5) is equal to $N \times I_{DS}^1$, with I_{DS}^1 the device unit cell current and that other device parameters, such as threshold voltage, sub-threshold slope etc. remain the same. This shows that the periodic boundary condition of the simulator is satisfied.

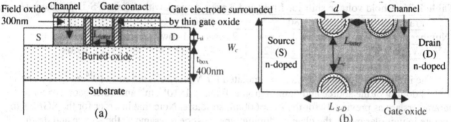

Figure 3. (a) Side and (b) top cross-sectional view of a SGFET unit cell for simulations. W_c channel width, L_c inter-gate distance within 1 row, L_{inter} inter-gate distance between source and drain sided row, d gate diameter.

RESULTS AND DISCUSSION

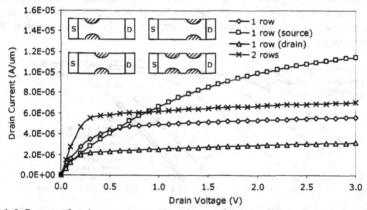

Figure 4. Influence of gating row arrangement on output characteristic (L_{S-D}=600nm). Inset: gate configurations.

The influence of the number of gating rows and their arrangements has been investigated in four different arrangements (Figure 4 inset). Structures are L_{S-D}=600nm, W_c=130nm, d=40nm,

t_{ox}=10nm gate oxide thickness, L_c=30nm, L_{inter} =200nm, t_{Si}=40nm, N_D=1x10^{14}cm^{-3} channel doping and N_D=1x10^{17}cm^{-3} ohmic contact doping. Results (Figure 4) show that the introduction of an extra row into a 2-row gating structure gives significantly better output characteristic (g_d).

In our proposed SGFET structure, screen-grid (SG) refers to the extra row of cylindrical gates added near the drain, in order to screen the effect of the drain voltage on the source-channel potential barrier. This 2nd gate row controls DIBL and thus reduces output conductance g_d in saturation mode. Table I shows the values of threshold voltage variation (ΔV_T) caused by DIBL calculated for both the 1 row and 2 rows structures for three deep sub-micron values of L_{S-D}: 80nm, 70nm and 60nm. In all the structures, geometrical dimensions are: W_c=50nm, d=10nm, t_{ox}=5nm, L_c=30nm, L_{inter}=30nm and t_{Si}=40nm. We observed that the 2 rows SGFET gives significant lower threshold voltage shifts in all devices, providing screening from the short channel effect.

Table I. Threshold voltage shift for 1 row and 2 rows devices with different S-D distances (L_{S-D}).

L_{S-D} (nm)	80		70		60	
Number of rows	2 rows	1 row	2 rows	1 row	2 rows	1 row
$\Delta V_T/\Delta V_D$(mV/V)	-66.3	-142.7	-74.4	-182.4	-84.95	-207.08

The influence of channel doping is simulated for N_D=1x10^{14}cm^{-3} to 1x10^{18}cm^{-3}. The ohmic contact doping of the source and drain region is fixed at 1x10^{19}cm^{-3} and other geometrical parameters are as previously for the L_{S-D}=600nm structure. Note that in order for the SGFET to operate in unipolar mode, the channel doping type must be the same as the source and drain ohmic contact doping type. The plot in Figure 5 shows the transfer characteristics of the devices in the sub-threshold region for the range of chosen channel doping densities. The sub-threshold slope improves as the doping concentration decreases and reaches the theoretical minimum for an undoped channel. Therefore, our proposed structure works well with a lowly doped active region avoiding mobility degradation due to ionised impurity scattering. In addition, it avoids the fluctuation of the threshold voltage due to slight random fluctuations of channel dopant distribution especially in ultra thin or small dimensional devices [6]. This behaviour is identical to finFET technology.

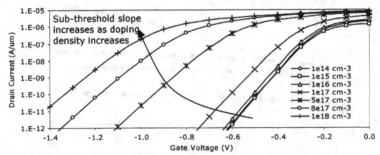

Figure 5. Drain current (logarithmic scale) as a function of the gate voltage shows the influence of the channel doping concentrations on sub-threshold slope.

Depletion mode SGFETs as those presented above (V_T<0) are useful for low power analogue applications but not practical for digital CMOS applications. In order to make the SGFET appropriate for digital applications, enhancement mode p and n-type SGFETs need to be made. The type of the majority carriers is determined by the ohmic contact doping in an undoped channel device while V_T can be controlled via an appropriate choice of gate metals (similar to finFET technology). V_T of an n-channel SGFET as a function of gate electrode work function is given in Figure 6, illustrating both enhancement or depletion mode operation of the SGFET with an appropriate choice of gate metal work function.

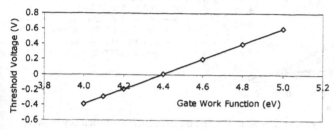

The influence of the thickness of the SOI Si body is also studied at t_{Si}=40nm and 200nm for L_{S-D}=600nm for both an SOI MOSFET and an SGFET, with similar geometrical parameters. Devices with a long channel were compared to avoid 2-D coupling between the source and drain electrostatic field

Figure 6. Influence of gate metal work function on threshold voltage, V_T of the SGFET.

distributions, ensuring that control is based solely on gating action. Figure 7 shows that for SOI MOSFETs, the sub-threshold slope, S and g_d degraded significantly as t_{Si} increases (see FD vs PDSOI) whereas for SGFETs, S and g_d are conserved as a function of t_{Si}. Therefore in order to increase the current through the SGFET two approaches are available without loss of gate control: a) increase the number of device unit cells (N) or b) increase the SOI body thickness (t_{Si}). This channel thickness independency is unlike finFETs, where an increasing body thickness or fin height rapidly deteriorates the performance in terms of S [7].

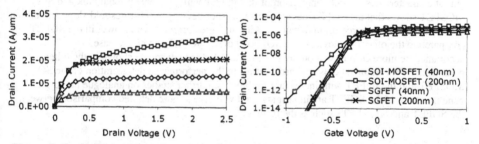

Figure 7. Comparison of the influence of channel thickness, t_{Si} (40 & 200nm) on the output and transfer characteristics of the SOI MOSFET and SGFET.

Finally, a comparison between the DC performance of an SGFET, a bulk MOSFET and an SOI MOSFET with similar dimensions, is presented. This study provides a rough idea of the SGFET performance as compared to its counterparts without optimisations. Simulations were carried out for both L_{S-D}=600nm and 60nm with t_{ox}=10nm and 5nm respectively. An SOI wafer of t_{Si}=40nm was used for both SGFET and SOI MOSFET. The channel doping density is N_D=1x10^{14}cm^{-3} and source and drain regions were doped at N_D=1x10^{17}cm^{-3}. Results obtained are given in Figure 8 and show that the SGFET offers the best sub-threshold characteristics and transconductance efficiency (g_m/I_{ds}) at short channel lengths (L_{S-D}=60nm) and comparable performance as the SOI MOSFET at longer channel lengths ($L_{S\text{-}D}$=600nm). Decreasing L_c whilst maintaining $L_{S\text{-}D}$ improves the performance of the SGFET even further and ensures that the SGFET maintains its performance advantage over the SOI MOSFET also at longer $L_{S\text{-}D}$.

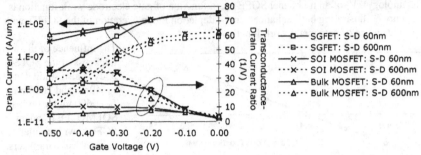

Figure 8: Comparison of sub-threshold characteristic and transconductance efficiency between SGFET, SOI MOSFET and bulk MOSFET.

CONCLUSIONS

This study presents TCAD (Taurus[TM]) results of a novel 3D FET – the screen-grid FET (SGFET) – where the gating action is due to cylindrical gate fingers inside the channel region perpendicular to the current direction. The novel gate structure ensures a de-coupling of the gate length and the source-drain distance, such that the transconductance can be optimised without degrading the output conductance. The 3-D simulations show optimal sub-threshold slope for undoped channels, near zero output conductance and good DIBL control for short source-drain distances, illustrating that the SGFET offers the possibility of downscaling without degrading the output characteristics. Direct comparison of the SGFET with the conventional bulk and SOI MOSFET shows that the SGFET outperforms the other devices in sub-threshold slope and output conductance at low source-drain distances. Although the current drive is lower in the SGFET compared to the other two devices, the SOI body of the SGFET can be increased to accommodate more current without loss of the performance advantage shown in the other parameters. This feature stands in sharp contrast to the performance deterioration of the SOI MOSFET and the finFET with increasing fin heights. Further studies on analogue and digital benchmarking of the SGFET are in progress and will give a more detailed comparison between the SGFET and other FETs such as the finFET.

REFERENCES

1. S. Asai and Y. Wada, Proc. IEEE **85**, 505 (1997).
2. *Taurus*[TM]. Version X-2005.10, Synopsys, Inc.
3. C. Bozler, G. Alley, IEEE Trans. Electron. Devices **ED-27** (6), 1128 (1980).
4. T. Langdo, et al., Solid State Electron. **48** (8), 1357 (2004).
5. *Executive Summary Of ITRS*, 2005, http://www.itrs.net/Common/2005ITRS/Home2005.
6. I. Polishchuk, and C. Hu, Appl. Phys. Lett. **76** (14), 1938 (2000).
7. G. Pei, et. al., IEEE Trans. Electron. Devices **49** (8), 1411 (2002).

Mater. Res. Soc. Symp. Proc. Vol. 913 © 2006 Materials Research Society   0913-D05-09

TCAD Modeling and Simulation of Sub-100nm Gate Length Silicon and GaN Based SOI MOSFETs

Lei Ma[1], Yawei Jin[1], Chang Zeng[1], Krishnanshu Dandu[1], Mark Johnson[2], and Doug William Barlage[1]

[1]Department of Electrical and Computer Engineering, North Carolina State University, Raleigh, NC, 27695

[2]Department of Material Science and Engineering, North Carolina State University, Raleigh, NC, 27695

ABSTRACT

Sub-100nm gate length silicon and GaN based SOI n-type MOSFET are modeled and simulated using ISE-TCAD (now synopsys_sentaurus). Several silicon SOI structures such as planar fully depleted SOI, FinFET, Tri-Gate MOSFET, cylindrical channel (OMFET) and triangular channel MOSFETs have been studied to compare the structure dependence of the device performance. Silicon and GaN as channel materials are also compared for these different SOI structures for projecting the device performance for very short channel SOI MOSFETs. Our study shows that for sub-100nm gate length, GaN based transistors have better Ion/Ioff ratio and higher small signal transconductance than silicon based transistors. And GaN and Si based devices have comparable performance such as sub-threshold slope and threshold roll off, etc. However for sub 20nm gate length, simulation shows that while it is not satisfying for silicon based device for digital applications, GaN based transistors with lower off state leakage current, less short channel effect than Silicon based transistors are still good candidates for digital applications . The TCAD study shows that GaN could be a promising candidate for making very short channel device as the GaN processing technology is advancing.

INTRODUCTION

Silicon MOS transistors have been scaled down fro over 30 years, and the gate length has reached sub-90nm in products and 5nm in the research level. Several different silicon on insulator (SOI) structures such as FinFET [1, 2], Trigate [3, 4], and Omega-field-effect transistor (OFET) [5, 6] devices have been receiving attention as potential device candidate for nanoscale silicon MOSFET. The FinFET structure we are going to study is one type of vertical double gate SOI structure, the channel is surrounded by the two thin layer of oxide as gate oxide and the top thick layer of insulator as shown in Fig. 1. The TriGate structure is also called triple gate structure looks like FinFET but the top gate oxide is also very thin so the channel is surrounded by three oxide layers as shown in Fig. 2. The Omega-FET is the structure where the channel can be further wrapped around by the gate oxide layer, the extreme case of the OFET structure is the cylindrical channel fully wrapped around by the gate oxide as shown in Fig. 3. The triangular channel MOSFET structure is on kind of double gate SOI structure, but the cross section view of the channel is not in a traditional rectangular shape but in the triangular shape as shown in Fig. 4. All of the above mentioned SOI structures have been proposed and the simulation work has been done individually. However the structure comparison has not yet been done for all these devices for device optimization.

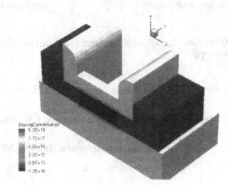

Figure 1. The FinFET MOSFET structure

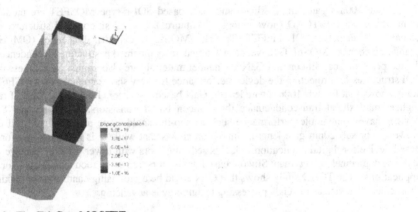

Figure 2. The Tri-Gate MOSFET structure

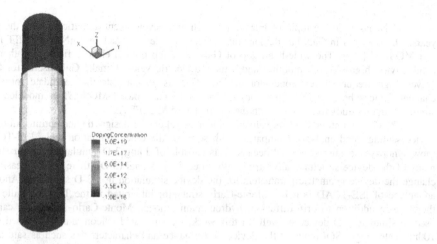

Figure 3. The OMEGA Gate MOSFET structure

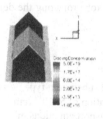

Figure 4.The Triangular channel MOSFET structure

GaN has been the topic of intense research and development activities during the past years. The progress in GaN has demonstrated successful devices such as GaN based HFET,[7] and MOSFET.[8, 9] The wide-band gap of GaN, the high critical electric field, high thermal conductivity, high electron mobility and saturated electric velocity make GaN useful for high power, high frequency semiconductor devices. GaN is also a potential candidate for short channel devices because of large bandgap. Therefore GaN based MOSFET is modeled and simulated in this study to compare with the silicon based MOSFET.

TCAD simulation tools are valuable aids in novel device design, device optimization and device scaling trend analysis. Comparing with some of the other simulation tools, ISE-TCAD (now synopsys_sentaurus) is used because it is capable of doing 3-D simulation. In our study, many of the device structures are intrinsically three dimensional (3-D), therefore it is easier to change the device geometric parameters for the device simulation in a 3-D simulator. Another advantage of ISE-TCAD is it has Monte-Carlo simulator integrated in the TCAD bundle and therefore in addition to drift-diffusion, hydrodynamic model, Monte-Carlo simulation can be used to simulate 2-D device as well. In this work, we simulated silicon and GaN based sub-100nm gate length SOI devices, the device performance and characteristics such as saturation current, off state leakage current, threshold voltage, sub-threshold swing and drain induced barrier lowering (DIBL) are compared through the above SOI structures. Silicon and GaN as channel materials are also compared for these different SOI structures for projecting the device performance for very short channel SOI MOSFETs.

EXPERIMENTAL DETAILS

For both silicon and GaN based devices we have studied, we use substrate p-type doping concentration of 1×10^{16} (cm^{-3}) and source drain n-type doping concentration of 5×10^{19} (cm^{-3}). The gate oxide thickness is set as 1nm. The Hydrodynamic mode of transport equations in DESSIS of ISE-TCAD are used to account for energy transport of carriers. The device self heating effect is also modeled by including the thermal-dynamic model. To study the inversion layer quantization effect, the Quantum model 1-D density gradient model is used. For Si based MOSFET, we use the Philips unified mobility model because it is well calibrated and widely used for MOS devices. For GaN based MOSFET, we use Arora mobility model, with Canali model as high field saturation model to account for the high saturation field and velocity of GaN bulk material.

DISCUSSION

Silicon SOI MOSFETs

The silicon based SOI devices are simulated, the following Figure 5 shows the Id-Vg curves at Vds =1V of several SOI MOSFETs with the gate length of 45nm. The multiple SOI structures have all shown better sub-threshold slope than planar non-depleted MOSFET. The Triangular channel MOSFET from the simulation seems also a good candidate for better control of the sub-threshold leakage current. One of the figure of merit (FOM) of digital device is Ion/Ioff ratio, as seen in Figure 6, the Triangular channel MOSFETs simulated have demonstrated larger Ion/Ioff ratio for above 20nm gate length devices. All MOSFETs show very low Ion/Ioff ratio as device scales down to sub-20nm. Quantum effects have to be taken in to account for accurately simulation of device performance beyond this 20nm node. Figure 7 shows the threshold voltage roll off due to device scaling. As shown here, the multiple gate devices have the benefit of better control the threshold voltage roll off and less device performance variations due to processing variations. Triangular channel MOSFETs also demonstrated comparable threshold roll off among the other SOI structures. Figure 8 shows the transconductance of triangular channel MOSFET is lowest among all the MOSFET structures.

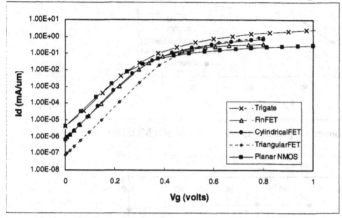

Figure 5.The Id vs.Vg for gate length of 45nm MOSFETs

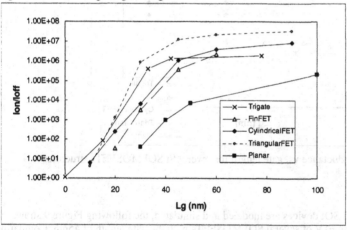

Figure 6.The Ion/Ioff ratio vs. gate length for several Si SOI MOSFETs structures

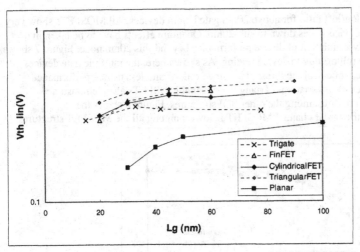

Figure 7.The threshold voltage vs. gate length for several Si SOI MOSFETs structures

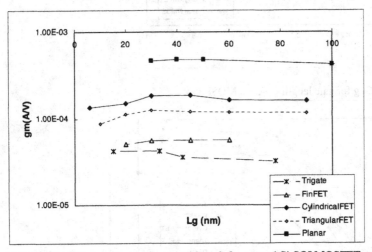

Figure 8.The Transconductance vs. gate length for several Si SOI MOSFETs structures

GaN SOI MOSFETs

The GaN based SOI devices are modeled and simulated, the following Figure 9 shows the Id-Vg curves at Vds =1V of several SOI MOSFETs with the gate length of 45nm. Comparing the multiple silicon SOI structures and GaN based SOI structures. We find the GaN triangular channel MOSFET shows better sub-threshold slope and low off state leakage current. From figure 10, in general, for low dimensional devices, GaN based MOSFETs have all shown better

Ion/Ioff ratio than Si based MOSFETs. The lower Ioff is because GaN has larger band gap, lower intrinsic doping concentration and lower minority carrier concentration, all these lead to lower junction leakage current which is the major concern for short channel devices. GaN device has lower Ion than Si device for longer gate length, this is because here we assume the electron channel mobility of GaN can achieve 1000 cm^2/Vs which is less than Si electron channel mobility, but this is almost the highest achieved in GaN HFET, we believe as processing technology advances, electron mobility in GaN MOSFET can also increase. As device scales down, the junction leakage contribute more to Ioff than the mobility to Ion, therefore the FOM Ion/Ioff ratio of GaN is better than Si as we simulated. The threshold roll off and small signal transconductance of GaN based MOSFETs show similar characterization with silicon based MOSFETs, and therefore we are not going to show them in the paper but the GaN based MOSFETs show larger transonductance. Using Ion/Ioff = 10000 as a critical ratio for digital applications, beyond 20nm gate length, silicon based MOSFETs are not satisfying this requirement and GaN based MOSFETs with much better Ion/Ioff ratio are potential candidates for future generation digital devices.

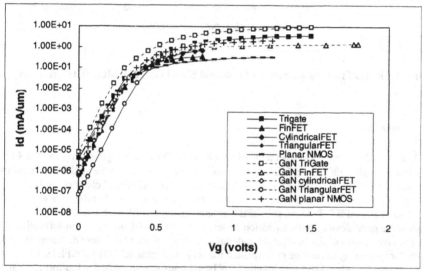

Figure 9.The Id vs. Vg for gate length of 45nm MOSFETs for several Si and GaN SOI MOSFETs structures

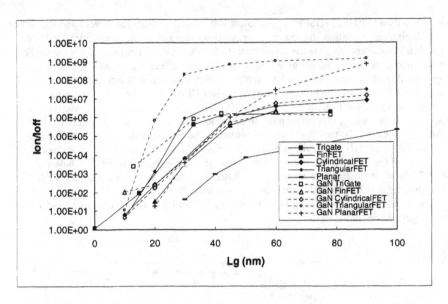

Figure 10.The Ion/Ioff ratio vs. gate length for several Si and GaN SOI MOSFETs structures

CONCLUSIONS

In this study we have modeled and simulated sub-100nm gate length silicon and GaN based SOI n-type MOSFET using ISE-TCAD (now synopsys_sentaurus). Several SOI structures such as planar fully depleted SOI, FinFET, Tri-Gate MOSFET, cylindrical channel (OMFET) and triangular channel MOSFETs have been compared. Apart from the other multiple gate MOSFET such as FinFET, TriGate, and OFET, triangular channel MOSFET has also demonstrated a good device characterization in device sub-threshold slope and Ion/Ioff ratio. Silicon and GaN as two different channel materials are also compared for these different SOI structures for projecting the device performance for very short channel SOI MOSFETs. Our study shows that for sub-100nm gate length, GaN based transistors have better Ion/Ioff ratio and higher small signal transconductance than silicon based transistors. And GaN and Si based devices have comparable performance such as sub-threshold slope and threshold roll off, etc. However for sub 20nm gate length, simulation shows that while it is not satisfying for silicon based device for digital applications, GaN based transistors with lower off state leakage current, less short channel effect than Silicon based transistors are still good candidates for digital applications . The TCAD study shows that combining with SOI structure optimization, GaN based SOI MOSFETs could be a promising candidate for future generation digital devices as GaN processing technology is advancing.

REFERENCES

1. H.-S.P.Wong, K.K.Chan, and Y. Taur, *IEDM Tech. Dig.*, 1977 pp. 427-430.
2. D.Hisamoto, W.-C. Lee, J. Kedzierski, H. Takeuchi, K. Asano, C. Kuo, R. Anderson, T.-J. King, J. Bokor, and C. Hu, *IEEE Trans. Electron Devices*, vol. 47, pp2320-2325, Dec. 2000.
3. B. S. Doyle, B. Boyanov, S. Datta, M. Doczy, S. Hareland, B. Jin, J. Kavalieros, T. Linton, R. Rios, and R. Chau, *VLSI Technology, 2003. Digest of Technical Papers. 2003 Symposium on* , 10-12 June 2003, Pages:133 – 134.
4. B. S. Doyle, S. Datta, M. Doczy, S. Hareland, B. Jin, J. Kavalieros, T. Linton, A. Murthy, R. Rios, and R. Chau, *IEEE Lett. Electron Devices*, vol. 24, No. 4, pp263-265, April. 2003.
5. H. Takato, K. Sunouchi, N. Okabe, A. Nitayama, K. Hieda, F. Horiguchi, and F. Masuoka, *IEDM Tech. Dig.*, 1988, pp.222-225.
6. S. Miyano, M. Hirose, and F. Masuoka, *IEEE Trans. Electron Devices*, vol. 39, no. 8, pp. 1876-1881, Aug. 1992.
7. M. Asif Khan, X. Hu, G. Sumin, A. Lunev, J. Yang, R. Gaska, and M. S. Shur, *IEEE Lett. Electron Devices*, vol. 21, No. 2, pp63-65, Feb. 2000.
8. K. Matocha, T. P. Chow, and R. J. Gutmann, *IEEE Trans. Electron Devices*, vol. 52, No. 1, pp6-10, Jan. 2005.
9. F.Ren, J. M. Kuo, M. Hong, W. S. Hobson, J. R. Lothian, J. Lin, H. S. Tsai, J. P. Mannaerts, J. Kwo, S. N.G. Chu, Y. K. Chen, and A. Y. Cho, *IEEE Electron Device Letters*, vol 19, No.8, Aug. 1998.

AUTHOR INDEX

SUBJECT INDEX

Printed in the United States
By Bookmasters